DON'T CALL IT DIRT!!

by
Gordon Baker Lloyd

BOOKWORM PUBLISHING COMPANY

Published by
Bookworm Publishing Company
P.O. Box 655
Ontario, California 91761
ISBN: 0-916-302-12-1 Hard Cover
ISBN: 0-916-302-04-0
Credits:
Cover and Book design by Hal Arnold
Composition by Tech-Art Services
Photos by Lee Simms, Sierra Madre, California
Text illustrations by the author
Photo model: Judy Latham, Author's daughter-in-law.

DEDICATION

To my mother, now resting, who got me started in realizing the good earth . . . her interest in wild flowers, her classes and her books, kept me on my toes . . . to my father, at this writing active at 96, who encouraged me since I was 19 to do what I wanted to do—teach, lecture and write horticulture . . . to my thousands of workshop members who said to "do it" . . . to Ken and Pat Kraft my favorite gardening authors, who made me feel that any active gardener may have something to say from actual experiences.

To my wife, Sassy, who put up with the mess of typing, listening, reading, and encouraging—and who loves gardens and gardening more than anyone I know.

Thanks to Bert Bertolero and his son Buzz, who read the first chapters to see if this was the home gardening slant needed . . . (the Navlet's have one of the most respected group of nurseries in Northern California) . . . thanks to Thomas Eltzroth, Assoc. Professor of the Ornamental Horticulture Department, Cal Poly in San Luis Obispo, California—he reviewed chapters flying back and forth from the East! . . .

I would like to have done a short one like Wm. Cobbett, 1821, who wrote a guide for a gardener on Long Island to carry in his back pocket . . . and he wrote to the point . . . long since gone, but William if you could, maybe so could I! . . . this one is to you for sure!

. . . And this is dedicated to all you who read it . . . beginning or advanced . . . with the hope it will open a new vista as you peer into the good earth . . . and you can be sure it won't be the last short one on this so important subject . . . good reading . . . good practicing!

Your gardener on the grow,

GBL

PREFACE

When I was asked to do this "text" I felt that it should be done with one thought in mind. Keep it simple. We have enough complicated materials on soils. For that reason some of you will say "it did not go far enough" . . . "it didn't cover enough detail" . . . you are both right. I did it that way on purpose. I have had over the last 46 years, thousands in my classes and workshops to say nothing about the millions on radio and the 15 years on television, beginning and advanced gardeners, who wanted things said and explained simply. I can get just as technical as any research writer—but would it help you?

I admit that much of the material I refer to is found here and in the West, but the basics can be applied to anyone anywhere who is ready to work his or her soil. The greatest group of year around gardeners are found in the 11 Western states, and hence my remarks are given towards them. There is no use spreading yourself too thin. You can't be an authority on everything for everyone!

I started out with references and books you could read to follow up the many items. The first readers said great, "but where can we find the books?" . . . and some even put it bluntly, so what? "If you don't know what you are saying by now, quit!" . . . so, no appendix . . . I was tempted! My hundreds of old volumes I love to quote for they put it so plain and clear.

This is the way I garden. This is the way I teach. It is the way I write. Thanks for reading and I hope I can meet you in person someday shortly.

Gordon Baker Lloyd
Studio House, Sierra Madre, CA 91024
March 23, 1976

CONTENTS

PART ONE

"What Is Good Soil?"

Chapter One

LET'S LOOK AT OUR "GOOD EARTH"

Everyone of us want to know, very simply, what to do with our own ground, soil, earth, in our backyards . . . our half acre . . . and even larger areas of our own earth. We would like to improve our soil. This improvement we have called "soil conditioning" . . . and frankly I would like to call it *soil correction.*

Our many plant problems:

Most of our plant problems have to do with the soils in which they grow. I think we should take a good look at our soil and realize what we have beneath our eye-seen surface. I would like to start with this thought. This quote is taken from a book published in 1733 called the "Horse Hoeing Husbandry". It is worded in the very old English. It comes right to the point.

> "Since the most immediate use of agriculture . . . relates to their roots, they ought to be treated of in the first place."

Whether roots grow in a lawn eight inches deep, annuals two feet deep, or shade trees beyond ten feet deep; how they grow, depth or in spread, depends upon the earth in which they grow. Since you and I are growing many kinds of plants, we must be conscious of the soil in which they grow—and hope they prosper. Why does a plant in one soil grow better than in another kind of soil?

Let's look at soil types:

It is time you take a good look at the earth on which you walk. It is not just what you think it is. Sure, you kick it and it's hard or stiff, or it fluffs up in dust. You may live in a desert area and you are tired of sand. You may live in a gravel wash and it drives you mad. You can be in adobe that sticks to your shoes every time you cross it. Maybe you are in a "tract" condition where the soil has been scooped off from one area to another to make the fill come out level. Very few of us live in what would be called the ideal earth condition . . . or as home gardeners want to say, a "perfect soil". We look at our earth and try to do something to make it "work better". This working has to do, as far as we are concerned, in making our soil respond to what we think is a physical condition. We want to make our clay-adobe turn better without sticking to our shovels. We want our sandy soils to hold some water so that it is not necessary to water every day or so. This is our normal approach to our earth, our soil, our dirt.

We look at soils in nature:

You stand in the rain forest on the Olympia Peninsula; you look at the floor under the trees at the edge of Crater Lake; you gaze out over the lush green of the Mississippi Delta, or the wooded slopes of the Wisconsin Reds, or the green mountains of Vermont . . . or wherever you pause long enough to look. You observe first that there is a layer of fresh leaves. These are yellow to brown and can be distinctly seen in their character. Below this fresh fall of leaves are those that are beginning to "break down", and their characteristics cannot be seen. Underneath these is still another layer of what we so often call leaf mold. The materials have almost completely "broken down". Grass, leaf and twig are in a condition of very light materials. Under this layer we see the soil. What we do not see is the fine line, black in color, of something called humus. This is what is "broken down" by bacteria into plant food, and with moisture, thus the plant, grass shrub or tree is fed.

This is the way nature does it, and has been doing it since the beginning of time. This is the ideal situation. No digging, no cultivation. Just letting time do its job, along with nature's aids. It is obvious that this ideal situation we cannot have in most parts of our garden or our land. We are going to grow things in a concentrated condition. We grow things over and over in small areas. It would be impractical to have the litter and accumulation in the average backyard or growing area remain. However, what we can do is learn some lessons from nature and adapt them into our own various problems. We must learn how soil is built and how it works. We cannot take it for granted. We cannot neglect it. In doing what we want to do in improving soil, we cannot go "way out" either. There is a happy medium to attain and this is what our book is all about.

How our soils work:

Before we begin to take our individual soils apart in relationship to our own problem, let's look at how healthy soils work. The tendency is to make this explanation very complicated . . . and from the scientist point of view and from the view from the laboratory, these few paragraphs can fill volumes. And right that they should. But, in our anxiety to get our soils into what we want in good "growing condition", we should make the explanation simple.

The Link in the chain:

Here are the parts that make up our soil. The actual backbone, mineral matter; the organic matter in the soil, and the final result of the breakdown of organic matter, into humus air needed so badly for roots and life; and moisture which makes up the sap system of plants and dilutes the plant food and keeps bacteria alive. These are the basics. When one of these links is broken, the chain does not work. Something goes wrong. We know what happens when we over water any plant. We know that sometimes applications of plant foods don't work and wonder why. We have been taught, as long as we can remember, that you must "cultivate" to break the crust of the soil, but why do we do it? These factors are all involved in this chain that makes good soil work, from the desert to our mountains, from our sand to our heavy adobe and clay soils.

How the plant grows, root wise:

We are concerned in this book with soils . . and not entirely with the effect of light, heat or cold on leaves or roots. Basically we must feed our plants. This may be organic or inorganic food. We want results. Here is what happens. Plant food that is organic or inorganic is attacked by bacteria. The bacteria live in relation to the amount of organic matter in the soil. They release

The reason for proper soil conditioning is to allow these roots to grow easily in all directions. This is the first part of eggplant roots that extend with hair roots 2 feet beyond these.

an acid; this "breaks down" the plant food (never mind the complicated explanation) and moisture in the right amount in the soil, dissolves the food and the plant absorbs it at the 100th inch end of each root. That is IT very simply.

The problem of why soils don't work:

We may say that they are "too sandy". Or someone says "my soil sticks together and I can't work it" . . . "I could make bricks out of my soil" . . . and so on. We are always concerned with the soil physically, but it goes way beyond this. In heavy soils we get too much moisture . . . in light soils we don't get enough. In heavy soils we don't get the air needed. In light soils we get too much. **This is what happens in the ground we walk on where the roots** must grow that eventually produce our green lawn, our blossoming camellias, our carrots, or our fruit. Our chain begins to break apart. It's time that we take a good look at the *basics of what makes up a good soil.* We may never attain the perfect soil, but we can bring the worst of our soils into a better condition so that our roots grow better—and the resultant better plants.

The word "organic" and "organic matter":

Organic matter, in the terms we can understand, is something that was once alive. This can be animal wastes, plant tissues. It can come from manure, compost, sawdust or rotted peat bogs.

In this period of time (1975-1976) we have had some rather radical things said about organic matter and the word "organic" has been badly abused. There seems to be few middle of the road discussions about it. There are exceptions in this study, and I must refer to two right now for your reading:

Sunset's Organic Gardening and *Growing Food the Natural Way* by two of my favorite, down to earth, home gardening authors, Ken and Pat Kraft.

We know that organic matter in the soil comes in two forms. One is the organic matter that is in the process of decomposition. The other is the actual humus which is the final end of organic matter breakdown. Going back into the very old books on farming, or early English books for home gardeners, it was readily admitted that you should put materials on the surface of the soil and dig them into the ground. Farmers were even advised to rotate soils. In Biblical times farmers were advised to let the ground rest every seven years.

Why the rest? Why the old precepts? It was simple. Weeds grow, plant materials accumulated; life in the soil came back to its normal condition on its own and plant food began to accumulate again. It is obvious we cannot do this now under our concentrated gardening conditions, but it is good to know why it was advised and how it worked.

Maybe the English went overboard in talking about soil conditioning with their deep trenching but was it wrong? We know that they were not. They were trying to build a good depth of soils for deeper roots and the use of organic matter was part of it.

Why we buy organic matter:

We buy sacked manure, peat moss, sawdust, to make our soils workable. This is the first and basic reason. So far you have had just a wee insight on other plus factors of organic matter. We buy these materials on the advice of our nurseryman or from other sources.

1. *We have heavy soil:* It may be clay or adobe or in combinations. All we know is that it is hard to work and turn over. It is sticky. We want something to make it workable. Water stands too long. So we use organic matter to break it up.

2. *We have sandy soils:* These don't hold water. They also pack. We want something to give them bulk. We add organic matter.

Results:

We find if we use it correctly and keep it up with its use, our soils will begin to work better. Some of our texture problems get solved and the garden responds to this treatment. This is what we are after and we are happy. As a reader of this book, I want you to learn the plus factors and know a little bit more of how to improve your soil and why you do what you do. When this happens, you will become a better soil manager and a better gardener or small scale farmer.

Can we over use organic matter? Yes, indeed. When you get up past certain percentages, you begin to defeat the entire soil conditioning program. To listen to some earnest people in this field you almost have to live on a six inch layer of the "stuff". Now think it out carefully, from what you see in nature and what you observe in the best gardens you visit.

How much by content?

We really don't know. We know that in most basic soil mixes, if we use organic matter by one third content of our soils, it breaks down to give us about the right amount of resultant humus. We must remember this breaking down process. The resultant amount of humus is small for sure. In container planting we may go as high as two-thirds by volume of different types of organic matter to one part of good soil.

I don't feel that I have the place in telling exactly how much to use. For your problems vary from garden to garden plot, from area to area. You will do it differently if you work in raised beds. You will do it differently if you work with pots or boxes. You will do it differently if you are working an acreage. All I can do is to give you some fundamentals—not from laboratory experiments but from just good gardening and watching what good gardeners have been doing during my 45 years of looking and peering all over the Western Area. I will try to give advice on some proportions to use that I find work . . . but even those can vary. So let's go further in finding out what makes good soil.

Chapter Two

NATURAL LIFE IN OUR SOILS

Wm. Lyon in 1904 in *Gardening in California,* said, "nature will have her way. Refuse to apply her laws and failure will follow; adopt her suggestions and a success can be obtained surpassing belief." One part of any soil discussion is the role that nature and living things play in our soils. We give little credit to bacteria, fungi and worms that surely help to make better soils. At the moment we won't discuss fungi, but we can talk about worms and bacteria to their soil action.

Speaking of worms:

This is a subject that has gone from one extreme to the other. To read some ads from enthusiastic worm growers, all you have to do is to "buy their worms" and you have the ultimate in soils. Worms are part of good soil—but let's not go overboard. It's interesting to note that as we improve soils, worms begin to be there in the natural way. Sure you can buy worms, but do it with some knowledge of how they grow—and for the specific purpose of your special need.

We have to go back to Darwin when he said, "It may be doubted if there are any other animals which have played so important a part in the history of the world as have these lowly creatures." He studied them until his death. He felt that worms pass as much as ten tons of dry earth per acre through their bodies each year and that amounts to about one inch of soil on the surface in five years. Our USDA has said that in some areas where they have done testing of earthworms working soil, they convert 700 pounds of soil into castings per acre per day. These castings are put into the top few inches of native soil and cause a build up of top soil. They even help in increasing water stability in soil from 3½ to 27 percent.

Chemical nutrients rise in soil where earthworms are growing. Nitrogen goes up 500 percent; 700 percent for available phosphorous; 1200 percent in exchangeable potassium and 150 percent in exchangeable calcium. This means availability to plants. All this is done by earthworms going back and forth from lower areas to the top surface. As they go back and forth they ingest and digest the organic matter available and leave what is called castings, like animal manures. These are rich in major and minor elements. The earthworms not only make narrow holes in the soil which help in air and moisture penetration, but they also leave behind a small amount of plant food. These are often gathered for potting mixes, soil amendments and so on. I suggest you read my reference suggestions for more information.

As soil improvers, we want to work with the creatures of nature who help our soil. They cannot work in a soil that has no organic matter. This is the reason why organic matter has such an important place in our soil discussions. It is to be noted that worms come into our soil when we begin to basically treat it right.

These are very timely references that I feel you can read with certainty of truth on the subject. *Earthworms for Ecology and Profit. Vol. 2; Earthworms and the Ecology.* These sell for $4.95 and are found at your bookstore or direct from Bookworm Publishing Co., 1207 Palmetto, Ontario, Ca 91761.

Speaking of bacteria:

This is one phase of soil conditioning that we talk little about, write little about and research for the backyard gardener has not been of the greatest! It is a subject that is argued about as to its importance. I believe your opinion depends on the background and knowledge of who says what, what you read, even what you are taught, and maybe even personal research. There are some things we do know.

What bacteria does:

Plant food does not enter into a plant as you apply it in the ground. Food elements in your soil are not readily available to your plants. Something happens first. The minerals in the soil, the food you apply, organic or inorganic, is broken down by several types of bacteria. They actually release an acid that breaks down the food. Bacteria live in the resultant end of the organic break down which is called humus. The story of bacteria, the kinds, is very complicated . . . and if you want to study them further you can do your own research. Suffice it to say, we know they are there and they do the work that is needed. Without them we would be lost in growing anything in the soil They are what break down composts. They are what break down the litter on the ground floor in the forest. Soils without them are sterile. It is that simple.

How many are needed?

The fungi and bacteria organisms are like miniature plants. They have to get energy someplace other than from sunshine. To get this energy they must live off organic matter. Great numbers exist in the soil and some bacteriologists estimate that there are as many as 46 billion in a gram of decomposing material. Around 100 billion per pound of soil or half to one ton per acre! Another authority says that their total weight, in a top foot of soil, is around a thousand pounds or about .03 percent weight of any good soil. All of this goes to prove they are there and badly needed.

My own experience in bacteria counts:

During World War Two, I worked with a top bacteriologist with his own laboratory. He took time out from his work to do some measuring of soil bacteria. We found in a tablespoon of soil we averaged around 15,000 of the good bacteria we needed while soils without organic matter came down to a count under 2,000. We did not work with sandy or sterile mixes. Just backyard soils. We even built a bottle of bacteria for compost inoculation—that's another story! The research we did was in outside home gardens and not in a laboratory. On our own we did find that the higher the counts of bacteria the better the health of the plants. The sick plants were always low in bacteria counts. We did several thousands of tests from the San Fernando Valley area, Santa Monica and Beverly Hills areas, giving us all types of soils.

We measured composts, we delved into manure masses. Everywhere we looked we found a direct relationship between bacteria counts and how plants grew. In summary we never found out which ones were the most important, but we did know that they came from one source, organic matter, either what is found in the soil, or what we add when soils are low in organic sources.

Chapter Three

AIR AND MOISTURE CONTENT OF OUR SOILS

Air in our soils:

In nearly all work with soils, it's easy to explain feeding and even watering, but air? Yet it is as important as water. 25 percent of soil in volume should be air. Some authorities say one third of the pore space of soil is air. As air is cut down in soil, things begin to happen. To keep bacteria alive, air must enter into the soil. With no air they begin to die. This is typical in heavy packed soils. In light soils, the air goes in too fast and there is too much. Organic matter, breaking down into humus, helps to get the right amount of air into the soil.

Roots and air:

We have found over the years the great importance of roots breathing. One reason we cultivated soil over the hundreds of years was not simply to cut down weeds or to hold moisture, but to let in air. Maybe we didn't realize it at the time we talked about cultivation. But we know now how important it is. Imagine my old 1733 book talking about how to cultivate and that if you "do good", things happen and if you don't other things happen that are harmful to plant growing. Over the past 50 years we have found out why this cultivation was so important in getting air to roots.

Examples on air in relation to plant food:

To reduce high alkaline soils we can use soil sulphur. Many a home gardener, and many a farmer takes soil sulphur and digs it in—6 or more inches . . . and then wonders why it doesn't work. We know that it must be in the top three inches or less of the soil so that sulphur bacteria can break it down. They must have air to do it. I know many home gardeners who dig in soil sulphur and wonder why they find it year after year when they redig their gardens. It has never broken down, but stayed inert. The ammonifiers require oxygen to change proteins to ammonia. The nitrifiers require air to change ammonia to nitrites then to nitrates to be absorbed into the plant. This all takes the right amount of air. One reason good soil conditioning and proper mulching is so important is to let in air. There are great many interesting studies on how roots breathe and often complete air circulation must take place in soil in order to keep plants growing.

Moisture in our soil:

25 percent of soil must be moisture. This must be moisture available to plants. Plants need moisture for a host of reasons. We talk about how and why much later. Look at fruit. How much moisture is an orange? Plum? Look at any stem and you find moisture, which we call sap. This is what not only dissolves plant food but carries it up into the entire leaf system. Here remarkable things take place. Bad gasses are absorbed and life giving oxygen is let out that keeps us living. Without moisture in the plant what would happen?

It is obvious to anyone who likes to grow things, no water: the plant dies. But in back of that take a good look. Moisture is needed to keep bacteria alive. No moisture = no bacteria = no life. No moisture = no absorption of food. Too much water in the soil the reverse happens the same way. Rot takes place, bacteria die, plant begins to die quickly.

It is no wonder that we can say that 80 percent of most plant troubles come from improper watering. Fruit split, bud drop, leaf yellowing, leaf drop, over-abundance of sap, even change of flavor and flower color can come from improper watering. Moisture is needed in the right amount to sprout seeds, make cuttings root, and on and on. We take it so much for granted. We use it and know it's needed, but are so careless in it's application. How precious it is not only to plants but to our own lives. How careful we should be with its use. We never appreciate moisture until we reach water shortages.

One of the important reasons in soil conditioning is to regulate the flow of moisture—its drainage or its holding qualities.

Chapter Four

FOOD CONTENT OF OUR SOILS

We cover this in detail in a later chapter. As to exact percentage of food needed in soil, mineral conent we might call it, will vary from authority to authority, and from text to text. It's confusing. It's no wonder we have so many formulas and so many different ideas.

In nature, plant food is there for the plant at all times. The fallen leaves rot down to humus, then into plant food which the plant can eat. There is mineral content in the soil, from rocks, that also breaks down as the proper action takes place. We need available mineral content by perhaps 3%. It is far less than other factors, but how important it is! Without food, the plant cannot live, the same as with us.

We have been taught that X element is the most important by far. We find that some elements are taken into a plant in greater proportion, but not necessarily is it the most important. We force plants, we want action. But in nature this never happens. It's a gradual process in feeding, small amounts frequently we might say.

In understanding plant feeding we go from one extreme to the other. From one claim to the other. The easiest thing to make and sell is plant food, dry or liquid. We are conscious of feeding. And this is proper. But take a good look at it in the eye of a naturalist who sees how important food is in the soil. Areas where food begins to be very hard to get plants, store it, adapt to it, and even die in part in order to keep the rest of itself living. Areas where we have lots of food and water and temperature that goes with it, look at the flowers and fruit we produce.

Since we do not have normal soils in any yard, we have to condition to get back to a more natural soil—and "food conditioning" is part of it. The earliest garden books dwelt at great length on food for soils. They had rather unique ideas on how to do it, but it all came basically down to one fact. Plants improve with feeding, and since some soils cannot do it, we do something so they have the elements needed.

PART TWO

TYPES OF SOILS

Chapter Five

HEAVY SOILS

Heavy soil . . .
particles stick together.

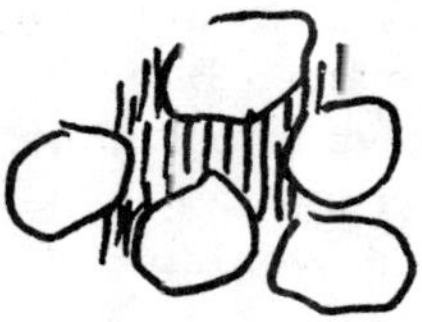

Organic matter in between soil particles opens soil.

The word heavy:

This has nothing to do with weight, but with the action it takes to work with adobe or clay soils. When we say "heavy" we mean that "blamed soil" that sticks to the shovel when we turn it . . . dries like a brick . . . gets gooey and sticky . . . and you name it! We pay no attention as to what is in it—just simply that it's the "worst soil we ever had!"

Adobe:

This is best known for its brick making capacity, used by Mexicans and Indians, in building. We know it as a grey to almost black and sticky soil that is hard to work. When it is broken down it is one of our richest soils Thousands of acres of vegetables are grown in this type of soil. Adobe, which is a home gardener's classification, is very high in clay particles which cause the sticky properties.

Clay:

This heavy soil may range in color from brown through yellow. It would be hard to miss true clay soil because of its color. The content of most soils are measured in ratio of clay particles. These particles are visible only through a very strong microscope and measure as little as 1/254,000 inch in diameter. Clay acts like a jelly or glue. Each clay particle has a moisture film which causes the stickiness you and I don't like. However, it must be remembered that some of this is very valuable for holding moisture and plant food.

How to make them workable:

We do this in two ways. First by natural means, and secondly by chemical means. The natural method is the addition of organic matter. Basically it is putting into the soil materials that break down into humus, which get in between the particles to give it better drainage, air circulation and as far as you and I are concerned, workability!

The use of humus forming materials:

Any kind of organic matter used to condition heavy soil must be mixed into the soil so that each particle of matter gets in between the clay or adobe particles. This means we must take a good look at the size of the particles we use and its actual mixing. Merely to put straw into the soil and calling it conditioning is nonsense. Home gardeners who dig leaves and compost into the soil and want physical results, will come up with no results at all. These materials must rot down before they begin to give any reaction. To get immediate physical reaction, we must use matter that is fine in content and has gone through some preparation so that it will break down into humus. The more we churn and mix these materials the better physical results we see at once. At least the soil is a bit more workable. In the meantime, the long range soil conditioning is taking place with the rotting of the organic matter into humus and its resultant bacteria action.

Chemical conditioners:

For years we have been using gypsum in alkaline areas, and lime in acid soil areas, to break up soil. The action is fast and we see results at once. What happens is the moisture layer between the particles is dried up, they fall apart, and our soil is workable. The folks coming West want to use lime as they always did in the garden and don't realize that they get into problems pH wise. They make the soil more alkaline than it already is in pH. Western gardeners have used gypsum for years for it does the same thing as lime, but the pH reaction is different due to the sulphur content. We have these materials in liquid and dry form. New, in the last five years, are products that have an entirely different reaction. They make water wetter and break down some of our worst heavy soils as well as hard pans.

If you use chemicals:

Whenever you use any of the chemical soil conditioners, be sure right behind their use you follow them up with organic matter. The conditioners available are biodegradable. This means, to you and me, they do not affect life in the soil . . . worms, bacteria, etc. These conditioners are short lived. Let's say the average is six months If you are going to spend the money on them to get the physical action started, let's make them last longer by adding organic matter right behind their use.

What heavy soil does in your garden—with no conditioning:

Fruit and shade trees: Large holes made in heavy soil with finest mix inside the hole, even two by two feet create a lot of problems. Roots cannot get out into the walls of the heavy soil as they grow in the unctuous mixture. Drainage does not take place. Roots actually rot in the presence of the fine mix in the confined area. This is like bringing up the young ones in a closet . . . little air . . . little drainage . . . no action to release plant food because of over amounts of moisture that cut down on bacterial action . . . and poor temperatures to make roots grow.

Solution: Create much larger overall areas, conditioned to allow further root expansion past a mere hole. This can be done by chemical means with addition of organic matter. If a unit area is needed for fruit trees, planting areas had better be 6 feet square and 3-4 feet deep.

Shrubs . . . roses, etc.: In heavy soil even 18-inch square holes bring problems. Most roses die from over-watering and poor drainage in their little hole than from poor quality. The larger hole than the can theory, gallon or five-gallon can, does not work, for again roots won't grow into hard walls of adobe or clay.

Solution: The larger the area that can be completely soil

conditioned, the better the results of the rose garden or the foundation plantings of our shrubs. This means overall conditioning of entire beds and special areas where we want to grow our favorite plants. . . . If you cannot condition an entire area, at least make holes three to four times the size of the ball or can size.

Lawns and ground cover areas: A root system of a good lawn should be at least 8 inches deep, and for ground covers 18 inches or more. How then can a little top soil, a little organic matter, put on the surface of a heavily packed base create a good root growing area several inches below? It simply doesn't work. For the first six months the lawn looks great, then things begin to happen. Water runs off. Spots get very bare. Grasses thin out. The lawn begins to die. As it goes down hill we water more, feed like made and wonder why it gets worse. Ground covers do well for a few months, begin to brown, and eventually thin out and die from the same reason.

Solution: Deep soil preparation must take place. You must work with the soil you have. Bringing in a contrasting soil does not work. There must be good drainage. The soil must be broken down so the little roots can grow. There are hundreds of plants per square inch of lawn. Since the lawn cannot be cultivated per se, we must build a soil that expands and contracts. Therefore, serious soil conditioning must be considered.

Vegetable garden: We have the same problem here as with other plants. Most vegetables have root systems of two feet or more. Even radishes and carrots have roots that go way beyond the typical part we eat. When soils are heavy, carrots split, root crops get tough; tomatoes split; corn grows very poor ears, and no crop ever reaches peak size.

Solution: Again, we must open up the soil. Since our crops are normally sown from seed, it is most important that the soil retain moisture and drain well to get seeds to sprout. The deeper rooted crops must have good drainage and freedom in which to move and special root crops must have room to swell out to normal size. Soil conditioning is very intensive. This is one reason many home gardeners are going to raised beds where intensive work can be done and drainage maintained at the highest level.

The flower beds: There is nothing worse than planting pansies out of a real good soil mix in their flat or container, into the heavy soil, and in a short time have them die. The same for any annual or perennial. What we have again is lack of room for the roots to grow. Drainage is nil and rot takes place. Often we are advised to raise flower beds for this one reason.

Solution: Better soil preparation for sure. The old English gardeners often took a full year to prepare their flower beds. Digging in compost and manures over and over. They knew that the job had to be done thoroughly in order to give room for roots and for a more permanent home for the roots.

Pot, planter and tub growing: Why do we have special mixes for containers? Very simple. The plant is growing in a concentrated area, somewhat abnormal, so we prepare mixes that not only hold moisture, but drain well. It is black and white. Either you do it and have a good plant, or don't and they die. There is no happy medium in container growing. In our outside ground areas we get careless and plants manage to get by at times, at times they don't, and we wonder why.

Chapter Six

LOAMY SOILS

What does loamy mean?

This is the soil we feel is ideal. Often we call it top soil. Actually there are many kinds of loamy soils. Those with more sand content are in contrast to those with high clay content. Normally these are soils that are 50 to 80 percent sand, 20 to 50 percent silt and about 2 to 4 percent organic matter. They drain well. They hold a fair amount of moisture and have some plant food.

How they act:

Since they are easy to work and seem to be more "open", they are easy to dig and to plant. They take the usual amount of organic matter with ease. They turn quickly. They don't seem to pack and cultivation is easy. They water easily. Moisture does not stand around and cause drainage problems.

How we use them:

If our soil is abnormally heavy or very light, or almost straight sand or gravel, and we buy top soil by the yard and apply it 3 to 4 inches deep over our horrible mess. And there it sits. It is a poor substitute for good soil conditioning for lawns, flower and vegetable beds. The material eventually disappears back into the present soil you have and you are right back where you started. However, it does have some use. Here is the limit of its use:

1. Use it to fill areas that need levelling. Even then work it into the soil as a mix. Or when it is used as a deep fill, be sure there is good drainage below it and into it mix copious amounts of organic matter. Join it with the other soil.
2. We use it in potting mixes and planter boxes, even raised beds. Here we have a different situation in which we plan to grow X plant in a confined area. We would like to have an "ideal soil" as part of the mix such as one-third soil, one-third sand, one-third organic matter, a typical planting mix and seed growing

mix in containers.

The natural loamy soil areas:

They are few and far between. You can remember the areas next to rivers. You can, no doubt, remember areas where vegetable gardens have had soil worked over so much that the soil is almost ideal. The mistake about top soils is that they are often merely soils taken out of fills, dug out of bare areas and have no relation to really good ideal soils. A few of us may be lucky to have great soil naturally. If we do, we can still improve it more by adding to it organic matter to give it more life and even more moisture holding content.

Chapter Seven

LIGHT SOILS

Light soils . . .
particles are
far apart.

organic
matter fills
between particles.

What is light soil?

This again has nothing to do with actual weight. In fact a gallon of sandy soil or gravelly soil might weigh more than heavy adobe! Light soils are soils that drain very rapidly, have little or no moisture-holding qualities and contain nearly no life due to lack of organic matter.

Sandy or gravelly soil:

They are from 80 to 90 percent sand, 5 to 20 percent clay and about 2 to 5 percent organic matter. Much of this type of soil is found in beach and desert areas, and some areas near rivers. Particles are 1/50-1/100th" in diameter.

More well drained soils:

We cannot call decomposed granites and gravel soils, sandy soils, but in home gardener terms they are certainly in the class of fast drained soils. They lack organic matter content and are very low in food value. Decomposed granite does pack as do the smaller sized granite soils. Hill areas have lots of decomposed granite soils and gravels are found in river and valley areas.

How to make them work:

The thing that makes these soils work is addition of organic matter. When it rots down, it gets in between the soil particles and fills them in. This acts as a sponge to hold water. It is obvious that some types of organic matter added to these soils at once begin to show a physical change. As the materials rot down they even act further in giving life to these soils that are nearly without any bacterial action. Decomposed granite begins to break down, moisture is held, and they become a fairly decent garden soil. Chemicals are not commonly used in the treatment of these soils unless they have hard pans below them that need breaking down.

How to use organic matter in light soil:

The trick here, as in heavy soils, is to get humus forming materials down in between each particle of soil. This means mixing. The more soil is mixed so the particles begin to get next to each other, the better the action. Since these soils tend to use organic matter faster, burn it up we say, they may have to be added more often till the saturation takes place and they get back to a normal condition.

What happens in light soil with no conditioning:

Fruit and shade trees: Roots run rampant in very light soils, close to the surface and great distances to get water. Water is not held around roots, and leaves tend to get lighter in color; fruit trees with poor crops; most of it due to lack of humus in the soil to release plant food or to hold it when it is put on.

Solution: In planting make large holes 4 to 6 feet deep and across with at least 25 percent organic matter in the entire root area. In established trees, heavy mulches and regular applications of organic matter may be needed till a good ratio of soil to humus is established.

Shrubs, roses: Most rosearians feel roses do best in heavy soil. Most shrubs would love a soil high in organic matter. The average garden shrubs want good drainage, but some confinement, so roots don't run wild. The exception of these average conditions are natives. They love fast drainage and a soil that is light.

Solution: Add organic matter as needed in overall bed planting and if hole by hole, make the holes 4 to 6 times the size of the container or ball.

Lawns and ground covers: Laws like a well-drained condition along with any ground cover. The problem is too rapid drainage and little life in the soil.

Solution: In both of these conditions you should add up to 30 percent organic matter to hold moisture and to give bacterial action.

Vegetable garden: Most vegetables like a light soil in which roots can go easily through the soil, especially the root crops, carrots, etc. On the other hand, it can be so light that their is no food value and the crop goes to leaf.

Solution: It is admitted that it's easier to work over light soil with organic matter than heavy soil. Copious amounts are needed to hold moisture, to give bacterial action to the soil. Mix thoroughly.

The flower beds: Any annual or perennial likes good drainage within reason. Just pure light soil does not produce bloom.

Solution: Add organic matter to give content to the soil, and to add the needed life and moisture holding qualities.

Containers and planters: You cannot fill a pot, container or planter with sand or light soil and expect it to work. That's no better than jamming a pot with adobe! In a container you would have a hard time getting moisture into all parts of it, to say nothing of holding moisture, even under normal circumstances . . . and on top of that there is no life in the soil to break down plant food. Half sand and half peat is great for rooting cuttings, but not growing plants.!

Solution: Add at least one-third organic matter and even another third of a good soil. This will give good drainage, let in air, and give life to the soil itself.

Chapter Eight

CROSS SECTION OF OUR SOILS

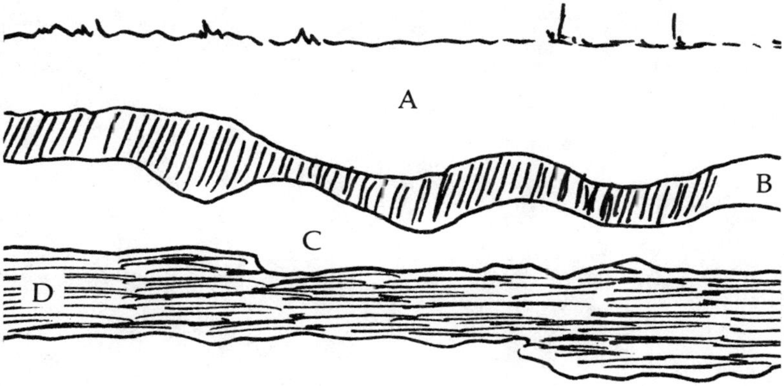

Soil Cross Section

A — Soil you see as you dig.
B — Layer of clay or rock.
C — Layer of sand, etc.
D — Layer of clay, rock, etc.

If you take a college course in agronomy, the study of soils, you not only learn about the three types we have talked about, but of all the soils by various names according to sections of state and country. Then you begin to study the actual cross sections of soil. Layers, sub-layers, and all that goes with it. Perhaps we have not given enough attention to this in our own particular condition. It is admitted that there is something below the typical soil we have. How deep does the soil we dig into go down? The top 12 inches? What's lower? What is worse? This entire thing may vary from yard to yard. So we cannot say this is the soil for everyone in a 4 square block area. So, some looking is needed if you are going to do the soil typing properly. Here is how:

Too often this is what happens before homes are built . . . materials are piled and dumped on vacant areas. These are graded, covered with earth. Here is a good before and after. This is why cross section study of your own soil is needed.

1. Dig the hole 4 to 6 feet deep. Wide enough so you can examine the sides of the hole to see the layers. A foot or so square should do it. It's a job, and that's why it's not done!
2. As you examine the layers, if any, take samples out into a tin can so you can examine it closely.
3. Mark the layers 1 to whatever layers you have. Label them at the depth found.

If you find a layer of gravel 2 feet down, this can add to lower drainage, but you will be concerned as to roots going through it. If the layer is a slab of rock, then you will have to worry about root penetration . . . and so on. In examining the layers and worrying about what to do with the sub-structure, you can get help from your nurseryman, and even from the Farm Advisor's Office in your County as to their suggestions in breaking layers. So often just a spadeful below the surface we run into what we call hardpan. Sometimes they are easy to be broken, sometimes harder. That's why good advice is essential as to what to use. In further chapters we will discuss how to do it. The main thing is to be knowledgeable of what you have soil wise, cross section wise in your own garden.

PART THREE

SOIL CONDITIONERS . . . SOIL CORRECTORS

Chapter Nine

WHAT ARE SOIL CONDITIONERS?

There are all kinds of packaged goods appearing each year that are touted as soil conditioners—those things that make your soil better physically. We have the greatest amount of them introduced in the West and, in particular, California. The reason is simple, for we garden here the year around. It is very easy to sell something quickly due to our many nurseries and garden centers. It's a great place for any company to introduce a product to see how it will "work"—and sell. Let's not start off this discussion by advocating organic or inorganic soil conditioners. Let's take a look at the basic ones we have on hand—those we can all find easily. I don't feel we should take up time to discuss rice hulls, coffee grounds, and any new special item that comes on the market. I am not inferring that they are no good; but I do feel we should relate ourselves to those items that are widely known and used. One small area may have grape pressings; another area may have rice hulls; another area some type of compost . . . and on and on. In your own situation, you can experiment with these items. I will lay down guide lines on this later.

What to watch for in new items:

Having been for many years on the introductory side of new things, I can speak from experience. Everyone is cashing in on anything organic. Some of it is good, some very far fetched. When a new item in soil conditioning is introduced, and no doubt we have some good ones coming along, you must be cautious. Buy for your soil just like you would for your kitchen appliances, shop tools, your car . . . and so on in your daily purchases. Who brought the item out? What is their background in the field? Is it a by-product they don't know what to do with and are going to try to sell it to you? What is the backing up of the item from your nurseryman? You can always ask about use tests and background of the product. If it can't be given to you, I say very bluntly, forget it. It is important that you know that the item is being introduced by a reliable company. It may be an old timer or a new one. But get some background. I am not going to pay $4.00 for a bag of something that I know nothing about—*unless it's value can be proven to me.*

"I've got" a great soil conditioner:

For sure you are not going to read this in any text. But I feel it's important in entering into any discussion on soil preparation. I have to refer here to some 45 years in the field and my personal involvement in introducing some very good items . . . and a couple that were real beauts!

When on radio or television, someone has to buy the time . . . and since we did a live show from our own garden on TV for over 12 years, we were a natural for a live demonstration. The time has to be paid for by someone and the temptation was great to sign a 13 week contract. (TV and radio are done in these blocks.) Soil conditioners were so easy to show and talk about. I was always a bit leary, but there were times when a big, reliable company told the station, "we are buying the time and that's it. . . ." I had a good average of saying no to some . . . but once or twice we bit!

In those years I talked to at least two dozen manufacturers who felt that they "had it made" . . . and some were weird for sure, and with good money to spend, it was hard to say no. Let's face it, if in your backyard, and yards of some neighbors and other friends, a product seems to work . . . It's hard to convince the fellow that wants to make his fortune that it may not work over a wide area under other conditions. Wide distribution, carrying of the product here and there, packaging, all enter into a product and its development. This you do not realize as you buy a product. There are many good items that have to be kept regional for this very reason. This is why I hesitate to talk about special "backyard" developments. Even now, I not only take a long look, but do my own very broad testing before I will recommend anything. What I am telling you, is that some of the highest promoted items in newspapers or other media, may not be the best investment for you. Take a good look with some reason at who is doing it, and what it will do for you.

Soil conditioning costs:

It may not be that the cheapest conditioner is the best. Sometimes the cheapest ones have the shortest life in the soil. The more expensive ones often last the longest and in the long run are your best investment. Many times a more expensive conditioner has, built into the cost: quality control (very important per bag or container), research and advertising costs. Briefly, it means reliability from month to month, as you may buy the item. A good established product has taken years to develop good commercial and home garden results. In talking about conditioners, I am not interested in the cost factor per se. I am interested in how long it lasts in our soils and it's results and benefits to our soils.

Dry or liquid foods?

You will have to work this out to your own situation. I tend to use liquid. In some instances it is better, such as container feeding. It is easier to use on ground covers and for use in irrigation. It may be that in some types of soils, side dressing of dry food will suit you the best. I can say this, loud and clear, let's not argue the fact. . . . I can add that some liquids do carry more minor elements and plus factors.

What are simples?

I have said that it is easier, and I feel far more practical to use a complete, all purpose food. But, we must understand we have what we call one element foods or simples. These are like superphosphates. This is pure phosphate and is used when we have very bad phosphate deficiencies. Ammonium sulphate is high nitrogen and dynamite to use. Bood meal is organic nigrogen . . . and so on. There is no reason to cover these here, for you find out by reading about all of them. To take them and mix them into a complete diet is not only hard, but expensive, and I hasten to add, some of them are very hard to get in many areas. I am surprised at some of the books that make up formulas so easy. I wonder where they have been recently. They certainly have not priced materials or have visited a broad spectrum of nurseries.

Organic or inorganic feeding?

This is a very tender subject for sure. My favorite writers, Ken and Pat Kraft, have done some very fine reporting on the subject. I think as an instructor with lots of actual gardening background, I try to hit the middle of the road. All plant foods, organic or inorganic enter into the plant the same way. Long tests of 25 years or more in plots with organic or with inorganic have shown no change in the mineral or vitamin content of the food crop. The use of organic matter in the soil tends to make for a better looking result—but that has nothing to do with the feeding per se. If you can find a plant food high in organic content, so much the better. It will last longer and not dissolve as quickly—and helps in building plant life. I am the first to avoid high analysis, potent, inorganic foods such as ammonium sulphate. Not only because they might repel soil life, but because they burn too easy and dissolve too quickly. I don't feel the subject is worth the arguments. I don't think it's worth the thousands of pages written on the matter. I feel that what we need to do is to begin some regular, decent feeding, with foods that have all the elements possible to obtain. I have had hundreds of vegetable gardens all over the country; I have worked with thousands of containers; hundreds of lawns . . . and to be very frank with you, I cannot clearly say you have got to "go organic or else" . . . I have stuck with this for 45 years teaching and lecturing . . . if I could see some definite proof, I would certainly change this paragraph for sure.

What about pH?

As we mentioned before, the control of pH is most important. We can help in its control by the use of plant foods on the acid side. This is for feeding everything in the garden. Some of our liquid foods are definitely acid in reaction. Nearly all the foods made in the West are built to combat alkalinity. For the plants that cannot stand any alkalinity—our shade plants like fuschias, begonias, azaleas, camillias, rhododendrons, etc., we have what we call acid food. The making of these foods were the first break to understanding pH correction. We now know that these same foods, in liquid form, are not used more frequently to feed everything: dawns, roses, vegetables, etc. As the years go by, in many areas, water's getting more alkaline. You are going to find all our foods in an acid pH range to control the alkaline problems. In very highly acid soils this problem would be in reverse, of course.

Chapter 10

SACKED MANURE

The present dry, bulk conditioners found in our nurseries and garden centers:

Taking a ten year period, five years ago and five years to come at this writing, I feel we will have these items to consider in any soil improvement. Let's start with the items found "bagged" or in "sacks". They are arranged in order of their "popularity". This has nothing to do with the best buy or how long they will last in the soil.

1. Sacked cow or steer manure. (In some areas you might find chicken and rabbit manures.)
2. Sludges.
3. Peats of all types. (Some true peat moss; others from peat soils.)
4. Leaf molds. These are very limited as to areas of distribution.
5. Wood products. These include sawdusts under several broad classifications.
6. Gypsum and related materials.

Sacked manures:

These are available from supermarkets, drug store parking lots to legitimate nursery sources. Prices have ranged over the years from 49¢ to $1.19 or even more. The amount in the bag can vary from 1½ cubic feet to 2 cubic feet. The home gardener pays no attention to the cubic feet in a bag. They see a sale on a given item and rush to buy it. After all, manure is the magic word, and in the background of any soil conditioning, this is true. Our forefathers—grandpa and even maybe our own fathers—dug into the earth what was available from nearby animals. I can ask, 'why not?' The farm yards with straw and manure dug into the soil brought results. Over the years, we have progressed far beyond the local contact. We now have manures bagged and ready for us to use in the garden.

Cow or steer manure, listed on the bag, has always been in great conflict. Bringing it down to the "promotion" factor, steer manure was considered as having been gathered from feeding lots or pens with cement bottoms, etc. These animals had a better diet for fattening reasons, and the results would be a higher content of food value in their droppings. The promotions on costs and values went on for years. One company against another promoted their sacked manures. At this writing, little is said about any food value. It is just there in price competition with home gardeners remembering the value of the word "manure".

As we look over the sacked materials now, we can find little food value. No longer is there an analysis on the bag. In the first place, there was little food value to start with a long time ago—maybe a little nitrogen and some other elements. It was used because it seemed to break down soil, for a bit at least, and plants seemed to grow better as the result of its use. This is understandable. What happened was that in the manures used, bacterial action took place in the soil, and plant foods in the soil became available. If other plant food was added, the manures helped to make them work still better.

As the years have gone by, we find that our sacked manures not only have less food value, but they have developed another problem. they contain salts that make our soils more alkaline and defeat food absorption. It is also doubtful that they last too long in any type of soil conditioning, heavy or light soils.

In some areas, various sacked manures are available. Perhaps, in a limited area, it might be rabbit manure; in another, chicken manure. These special manures can run higher in food content and even in bacteria, but as to soil conditioning value, one must take a long look as to what they do over a period of six or more months.

How to use:

The common sacked manures available at our nurseries and

garden centers are best used as top dressing covering seeds, like in a lawn renovation, or perhaps over a lawn in winter, spring and summer, as an additive to organic matter in the soil. They are seldom used as seeding or pot mixtures. In soil conditioning, at the rate of one to two inches on the surface and used in a reasonable manner, their longevity is very short . . . perhaps six weeks to four months at the most. Sacked manure has some value in the compost pile in adding bacteria. An inch of it in every three to six inches of compost, no doubt, can help.

Evaluation:

We are talking about the sacked manures you buy. These are, no doubt, in the long run, the poorest buy, not only in relation to longevity in the soil as an actual conditioner, but also the side effects of salts, etc., that affect all plants.

Past backgrounds of use of manures:

When you read the old books, from 1880 through 1930, on the use of manures, they were dealing with an entirely different material. It is completely different when you go to a barnyard and "get your own" mixed with straw and such. The materials are different and must not be confused with what we are talking about now. I can remember, in the thirties, being able to recommend a manure with an analysis and good body for conditioning. That is no longer the case now; so, as you read manure advocates, take a long look at their source and yours, for sure.

Commercial aspect:

It is true that at the moment the cheapest form of adding organic matter and life to the soil may be the use of some manure spreading over vast acres. Better that than nothing. I might disagree and advocate some other methods of helping soil organic wise, but at least the spreading of these products helps rather than hinders.

Chapter Eleven

SACKED SLUDGES

We have had commercial sludges for a long period of time in the garden market. Some over thirty years in the Western area. The first one was pebbled and imported from the East and sold more as a plant food than a soil amendment. Each area over the last fifty years has had little sludge promotions from one city or another, depending on its source. When we say sludge, we are talking about a digested sewage sludge that will come from the city sanitation system. The by-products go through a digestion system and composting that takes out any material harmful to humans in soil use. They have varied over the years for sure, and they have come and gone. The future of seeing more across the nation is bright, as new processes are discovered and cities realize the potential of materials they are throwing away. In this discussion, I am going to focus on the best known and distributed product of its type in the Western area. The material is called Nitrohumus and is sacked in 50-pound bags and can be purchased in bulk by the ton, which is about 2 to 2.2 cubic yards per ton. This particular sludge goes through a process that is not normally known and is being copied in other areas in the U.S. as the best process known at the moment. When it is taken from the hugh digesters, it is windrowed out on hugh, hard packed acres. In the long, high windrows, it is turned by special machinery daily to allow proper composting. When it has finally reached compost temperature of 140 degrees in the piles for a period of thirty-one days, it is now ready for packaging into the bags for nursery and garden center selling. Moisture control is very important, and in this process they can store and package at any time, rain or no rain.

How to use:

This is used as a natural organic fertilizer for turf, ground covers, flowers, vegetables, shrubs and trees. It can be used as an actual soil conditioner and very important amendment. It can be used as a seed bed cover over lawn areas, etc. It can be used on lawns, home and commercial, as a top dressing. Following directions on the bag, it can be used in soil mixes, in the ground and potting, in place of the use of any manure. It is highly recommended in bare root planting, vegetable bed preparation, and flower bed work, since there is no possible chance of any burn from ammonia. Use a 50 lb. bag for soil conditioning; a 2 lb. bag for 100 sq. ft. For top dressings, a 50 lb. bag covers around 500 sq. ft.

Evaluation:

I cannot overestimate the value of Nitrohumus. First, it is high in bacterial life—far more than any manure—without the burn factor and heavy odor, but with some food value. It even contains 2.6 percent iron. Anytime manure is used, I always suggest Nitrohumus—compost making, potting mixes, etc. In all soil conditioning, even the use of the finest wood materials, I always suggest the added use of Nitrohumus for the life it contains. This is why it's so fine for top dressing on lawns and all types of beds; in hanging baskets, pots and even small amounts in seed soil mixes. It has been found that it will increase germination rates and percentages of germination when used as directed.

We have talked about the addition of bacterial life to the soil. Here is a fast, practical way to get it. Many soil conditioners, soil amendments are dead to start with and only by breaking down can life be obtained. Here we can use this sludge with these other materials and begin at once with the proper amount of life.

Costs:

By the sack, on the average, we pay $2.79 for a 50 lb. bag. Compare the many added factors it has to manure. In the 50 lb. bag of manure, we pay an average of $.79 for what is called good quality. Longevity of Nitrohumus goes on and on with its bacterial quality an average of a year in duration. Action from manure stops at around three months. It burns out, as we have explained. Draw your own conclusions.

Brief History of Nitrohumus:

In examining any product, I think its history is so important. This is a good example of what I am talking about.

During the early 1920's, H. Clay Kellogg, Sr., while surveying the Santa Ana River bottom, observed the luxurient growth of vegetation along the river bank, resulting from sewage sludge, and took some sludge home with him to use in his yard. His yard responded similarly to the vegetation along the river bank and, as a result, drew many inquiries as to how he was able to create such a beautiful yard. It was at that time he decided to make a business of selling sewer sludge.

Making the product marketable in those days was quite a difficult task in that most of the processing was done by hand labor. The larger cakes of fertilizer, sometimes being one foot in diameter, had to be crumbled by the back handle of a shovel accompanied by two strong arms. Introducing sewage sludge to the "old time farmers" as a necessary part of their land preparation wasn't an easy task either, but in that results were what the farmers wanted and results were what they received, the business was showing definite progress.

During the year 1925, H. Clay Kellogg organized a fertilizer business, DBA Kellogg Supply Company. In 1927, Mr. Kellogg signed a contract to purchase and remove sewage sludge from drying beds of Los Angeles City and County agencies. During this period, the sewage sludge was pumped to very small lagoons where it was left to dry. When it was dry enough to handle, Kellogg used wagons and horses in the beds, where all of the material was picked up with pitchforks by hand and loaded on the wagons. This was the way material was brought to the screen and grinder for further processing. Because of the limited space, 80 acres of additional land was purchased to build large lagoons for drying the increasing amounts of sewage sludge. The sale of sludge was in bulk requiring only pulverizing for its acceptance as a fertilizer and was sold to the large agricultural area of first importance in Southern California.

As the years have moved by, they are now considered the leader in their field, becoming the largest supplier to landscape contractors and nurseries in the Southern California area. It is only fair to say they have complete lines of plant foods including many planting mixes and acidifiers.

To be technical, here is how Nitrohumus is made:

STOP I COUNTY SANITATION DISTRICT (2nd largest treatment plant in the U.S. and largest sludge composting plant in the entire world). This plant and the city plant (Hyperion) handle as much water as the Feather River Project.

Station 1. *Bar Rakes and Grinders* — The bar rakes remove coarse debris from the incoming sewage flow which would otherwise hamper treatment. Most of this debris (large pieces of organic material) is shredded by the grinders and returned to the flow. (Large items such as boards, logs, cans, etc., are trucked to the landfill.)

Station 2. *Grit Chambers* — 20 ft. deep — remove heavy inert material (sand, pea gravel, egg shells, etc.) that would otherwise cause undue wear on treatment equipment and would not be decomposed in the anaerobic digestion process. This material (grit) is also trucked to the landfill.

Station 3. *Sedimentation Tanks* — 12 ft. deep — remove the bulk of the solid material which will settle to the bottom or float. The floatable material is skimmed from the water surface and combined with the settled solids (sludge) which have been pumped from the tank bottom. This combined sludge (liquid in form) is pumped to the digesters. (From here the sewage, with solids removed, is pumped to the ocean.)

Station 4. *Digestion Tanks* — Sludge pumped from the sedimentation tanks is stabilized, over a 14-day period, in these large, completely enclosed structures using the anaerobic digestion process. This method employs bacterial decomposition of the sludge in the absence of oxygen and at a temperature of 95°F. The process destroys those pathogenic (disease causing) organisms which may be present, creates a burnable gas by-product and produces a stable liquid end product (digested sludge).

Station 5. *Generators* — Burn digester gas (60% methane, 40% carbon dioxide), supplying electrical energy to run motors and pumps for a large portion of the plant.

Station 6. *Boilers* — Burn digester gas to generate steam used in maintaining digesters at 95°F.

STOP II DEWATERING STATION — Digested sludge, still in liquid form, is pumped from the digesters to the centrifuges where the solids and liquid are separated. The liquid portion is returned to the plant for subsequent ocean disposal, while the dewatered solids (sludge) are discharged to a waiting truck, which dumps in on the field for composting.

STOP III COMPOSTING — The sludge is windrowed and composted under aerobic conditions for 14-15 days by daily turning and speedmixing, further drying and decomposing it. The 120-140°F temperatures attained in this process insure destruction of any remaining pathogenic organisms. Material is then picked up in 21 yd. self-loading carryalls for transporting to Nitrohumus stock pile or blending area for the making of Topper, Gromulch, pH Acidall, Potting Soil, Azalea Mix, Indoor Planting Mix, etc.

This shows there is more to a good product than meets the

eye. These pioneers are helping other cities to do the same thing and making these previously raw materials available to many more of you across the country.

Chapter Twelve

BALED OR BAGGED PEATS

Taking a fifty year period, we have had bagged peat. In the pre-war days, we had very dark and high grade quality peat from Germany. Then we began to get peat from Canada in huge bales. I can well remember the peat moss promotions forty years ago. After World War II we began to find some peats in the high mountain lake areas of California, and we began to bag so called peat from the Central Valley region in the Delta. The word peat had become very broad as far as the home gardener was concerned. He knew he could use it in light soils to hold water; and every book he read talked about peat in potting and seed mixes. Shade garden lovers always used peat in soil conditioning and container mixes. Anyone knew you could use peat moss in mulching. In heavy soil, it was widely used except it was a trifle hard to get into the heavy soils as a conditioner. Peat moss is now available everywhere, from little half pound bags at the five and dime store to the larger bags and bales at the nursery and garden center. Its value is for soil conditioning, mulching and as a help in correcting pH, although that is not the factor that it used to be. Food value is low. Commercially, it is frequently used in potting as an easy way for shipping plants in a light soil. With proper feeding, it's amazing what can grow in straight peat for some period of time.

Evaluation and kinds of peat:

Sphagnum peat moss: The one we know the most about is what is called sphagnum peat moss. At present it comes from Canada, bagged and in small bales. This is the light, fluffy stuff that we soak before we use. It holds lots of moisture. Its ash content is small—around 3 percent. It is, no doubt, best in potting and container mixtures. It is a good soil conditioner when thoroughly mixed, and, I add, never use dry! pH around 3.60.

Hypnum peat moss: This comes from lake bed areas. Its texture is nothing like sphagnum, and it varies greatly. It does contain some nitrogen. Its ash content is up to 30 percent, which means that as a soil conditioner it is less valuable than at first look. It is used in potting and container mixes and as top and seed dressings.

Sedge peat: This comes from the Delta areas and is dug right out of the ground. It can be brown to black and looks great! It is high in ash up to 50 percent. It is also high in salts and frequently may be so alkaline as to burn when sold in bulk from door to door! It will be the cheapest of the peats.

Costs:

A 2.5 lb. bag sells for $5.49. A 4 cubic foot bale of sphagnum peat moss, compressed as it is called, will run around $7.98. Bagged materials will vary in price. They are loosely packed. Two cubic feet will cover around 35 square feet in soil conditioning . . . much more, of course, when used as a top dressing over seed. pH runs from 5.0 to 6.0 in sphagnum peats.

Background:

Early California books in 1900 barely mention peat. As it became more available, in the period of 1920 and on, and promotions began to take place, home gardeners began to use it chiefly in mulching and potting mixes. Later it became widely used in actual soil conditioning. Sedge peats are of vintage to home gardeners after World War II.

Commercial aspects:

The sphagnum types are used at cutting mediums, potting and container mixes. They will always be used and have a very definite place to play in any commercial growing, both under glass, lath and wholesale growing of many plants.

Chapter Thirteen

LEAF MOLDS AND THEIR USES

As with peat, the use of leaf mold has always entered into potting and container planting and as a soil conditioner in shade garden planting. Old texts often refer to leaf mold, and it was gathered at that time from groves, sifted and used in mixes. Over the last fifty years, we have had every type of leaf mold you can imagine. Some have been so poor that they contained ground up fresh leaves. Others have been so excellent that they had to be kept to limited areas due to availability, weight and bagging problems. When we define leaf mold, we are thinking of materials gathered from a layer of leaves under the forest trees. The materials coming from half way through this layer to right against the earth—and trying not to get any of the earth! Leaf mold, at times, has carried the name of the tree from which it was gathered . . . oak leaf mold, redwood leaf mold, etc. In current years, we are not always able to keep this area so defined—so leaf mold is just what it says it is—completely broken down to partially broken down leaves. It may, at times, contain some small twigs. Poor quality contains large twigs, large leaves, etc. The finest leaf mold I ever saw came from the little town of Rescue in the Gold Country area and was under rigid control. Other good leaf molds have been scattered here and there. Current leaf molds seem to maintain good quality. The Lindauer people put out a top redwood leaf mold for many years, but when supplies ran out, they had to quit bagging the materials.

How to use:

Its best use, no doubt, is in planting mixes, pots, baskets, containers, etc. As an out and out soil conditioner in open areas, its purpose is defeated. Its value is to keep a mix open. It does help in drainage, and it does add some food value to a container or potting mix. Its ash content is around 30 percent. We frequently hear the proportion of one part good quality leaf mold, one part peat moss and one part of good soil for shade garden mixes in containers and raised beds. Bacterial life increases with its breaking down in the soil.

Costs:

A 2 cubic foot bag of good quality runs $3.49. That is why it is limited entirely to container mixes.

Past background:

Leaf mold is referred to in the earliest of texts, and, from time to time, slightly confused in discussions of piling up leaves and rotting them, which we call composting today.

Commercial aspect:

Many growers are getting their own sources of materials and piling and composting them to the quality they want. Leaf mold will always be used commercially for keeping certain mixes open.

Chapter 14

WOOD PRODUCTS

As we know the term now, it includes various sawdusts (chiefly redwood), materials from removing barks, and actual sizes of barks themselves, from either actual bark pieces to wood chips. The term then is very broad. In this book, we are concerned with materials for soil conditioning, and I will confine my discussion to the use of the sawdust group. There is really nothing new in the use of sawdust for soil conditioning. Books dating back to the 1600 period, and books in the early and late 1800 period talk about the use of sawdust in conditioning soils in farm and orchard work. It was found that this "waste" product, when used properly, lasted a long time in soil treatment. In heavy soils, it was at a high advantage due to its particle size. It would work at once to open up heavy soil. In light soils, opinions varied for it was felt that it took too long to break down to hold moisture. When it did, it was superb and might even be better than peat mosses.

Gardenwise, home gardeners often went to local lumber yards to get sawdust to use in their soils. It was free. Why not use it? When it was used, they found that for some reason or other, plants often turned yellow and died. Word got around that it was poisonous; therefore, don't use it. "Old timers", who evidently had been told from one generation to another, to add manure to the "stuff", had no problem. Believe it or not, the idea still lingers. When raw sawdust is used in any soil work, it takes nitrogen from the soil, which causes yellowing of leaves. When raw material is piled and composted for six months or more, this factor stops. We have found that we cannot only compost it, but to be on the safe side, we add nitrogen to give us the longest acting soil conditioner we have today.

As is true in anything good arriving on the market, there is always competition, and right that there should be. However, too frequently the competition is careless in quality control, in its anxiety to get a product to market. One of our best examples today is the use of redwood materials. We have redwood soil conditioners in all forms and in all word usages. For this reason, I am going to talk about the pioneer in the field of quality materials, just as I did in the Nitrohumus discussion.

When a good product is brought out, as I mentioned, now come the imitators. Now we have redwood sawdust, several dollars cheaper. Why? They even go so far as to color the product artificially and may mix nitrogen into the product dry, etc., so that from bag to bag or bale to bale, the quality varies so much that even burning has taken place. This has brought up some very rabid discussions on wood product use. Some have been very bitter, unfortunately.

As we look at the scene right now, we have the longest lasting soil conditioner, and one of the finest. When you want to break it down all the way through nature, what happens to tree trunks in the forest, anyhow?

Evaluation:

Looking at sawdust generally, we do have some fir sawdust in some areas. Ash content is around 2 percent. It may contain chips at times. Never use pine sawdust. It's problems are too complicated.

Commercial use:

There is nothing new in the use of sawdust in nearly all canned and boxed materials at the nursery. It's been going on for fifty years or more. In past years, even the use of shavings could be found in canned soils. It is becoming even more important in the growing of any canned stock at the growers' level. Just visit the "back area" of any large grower and be amazed. They may have their own direct source from a mill, but they pile it, compost it, and do it themselves, but they are using sawdust, and will, no doubt, use it even more in the future.

There are bark products, granular types, coming in all sizes that are used by orchid growers, and in special mixes. As I said before, we are concerned with you and me in the backyard soil problem. Redwood has about 1 percent ash content. It can be used in seed mixes, potting and container mixes, as top dressing over seed, in any soil conditioning from lawns to vegetables, from bare root planting to shade garden soil conditioning. I have used it in the place of peat moss for many years, and I notice its use is getting wider and wider. Its pH reaction, as it begins to rot, is better than peat moss. The immediate value is to break down heavy soil when it is mixed in between the soil particles. Physically, you can see it at once. When used in light soils, the same visual action can be seen. What is not discussed is what happens as it breaks down. This takes time, but when it does, just like in nature, we then have bacterial action, pH reaction, moisture holding qualities and, in the long run, some food value. Again, I add, this is not on the bag. This is not told to you at the nursery counter, but this is what happens. For that reason, good rotted redwood sawdust is one of the longest lasting soil conditioners, frequently continuing its good action over a period of one to four years. It is effective as a mulch, except in areas of high wind, as with many of the lighter types of soil conditioners and amendments.

Costs:

A 4 cubic foot bag, as an average, runs around $4.79. This will cover 200 square feet as a top dressing—100 square feet as a soil conditioner. This means, in the long run, the least expensive soil conditioner we have on the market, organic-wise with all the plus factors.

Background on Good Product:

A good product must maintain uniformity. In sawdust, this means size of particles and no variance of quality from sack to sack. The Lindauer Company of San Francisco, established around 1890, was working with wood product materials in use from tavern floors to cooking, believe it or not. In July of 1958, they decided that redwood sawdust, in great abundance and almost a waste material, could be used in soil conditioning.

After long research, they found that one of the factors in eventual garden use would be to compost and leech the material for a year in huge piles. The leeching would take out the tanic acid which causes some of the problems in garden use.

On top of the leeching and composting, it was decided to add nitrogen with very special equipment, just in case. To maintain an even texture, even quality control, new processes in mixing and bagging took place. As a result, the Big R we now have in the West is the ultimate in packaged materials, and, I hasten to add, they do put up their product for several high quality merchandisers, such as the Master Nurserymen in California.

As I said before, any good product has a background in its production and research. Here are some interesting facts on our leading redwood producer:

Background of Lindauer and their work:

Pine and fir sawdust was a necessary absorptive material, used on the barroom floors of the Barbary Coast in San Francisco. The Lindauer Sawdust Company had its humble beginning during this period. Butcher shops still use pine sawdust on the floor to prevent meat cutters from slipping on trimmed-off fats and gristles. Fir sawdust is used to absorb liquid spills such as paint, water, oil, etc. on ship docks, in paint plants, air terminals, warehouses, as well as for packing in shipping. The Lindauer people manufactured additional sawdust products such as oak, aler, redwood, ash and hickory for smoking ham, bacon, sausage and "hot dogs". Around 1955, redwood material was put to use on a large scale by many growers. The University of California had done extensive research into this wood specie as a soil substitute, mixed with sand, this producing the famous UC method of growing plants, without so-called soil. Redwood sawdust and sand were developed in various percentages for different plants. Lindauer did additional research on this new idea and, as a result, began manufacturing redwood soil conditioners for home gardeners throughout the Western states. The values of redwood, as opposed to other sawdusts, are the slow decay in the soil, the acid pH, and the discouraging of fungus and insects in the soil.

Chapter Fifteen

GYPSUM AND RELATED CHEMICAL CONDITIONERS

One of the most common of our chemical soil conditioners is gypsum magnesium sulphate It is easy to mix into soil and, as it gets in between heavy soil particles, it dries up the moisture layer and the particles fall part and the soil becomes workable. Soil conditioning with lime is very common in the areas of acid soils, and the results are the same. In the West, we use gypsum, which has some sulphur in it, giving it an acid reaction when used. There are several types of gypsum, from the kind used in plastering to the soil type that we use in the garden and farm. It is not as refined and does not need to be for soil work.

Gypsum, in some areas, has been dug right out of hillsides and put directly onto the land, such as in the gypsum diggings around the Bakersfield area. Farmers use gypsum regularly because of its small expense and easy use in spreading. Nearly all potato crops are grown on gyp-treated soils. There are also some liquid gypsums available in gallon bottles for home gardeners.

Amount used:

Normal use on heavy soil is around 5 pounds to the 100 square feet and churned in. The better you mix it, the better it works. When you have loosened the soil, be sure it is followed by organic matter to keep the soil open. Use your gypsum dry or liquid as a starter to better soil management.

Wetting agents:

The newest in our soil "breaker-uppers" are the liquid materials, one dry, that is used to make water wetter. We don't have to discuss the chemicals here. Read the bottle, but suffice it to say, under several trade names—the first called "Water-In", followed by "Water-Wet", "49er Soil Penetrant", "Germain's Pentrex", and "Break Thru", and many more. These are all super wetting agents that allow water to pass down through tight soil particles. Cost is reasonable; longevity from four to six months. Some have added minor elements and the needed three elements added to make a more all-purpose working material. Again, use them for what they are for. They get the soil started in breaking up particles. You follow with organic matter. They have a very needed use for slopes, for hard-packed lawns, for the breaking of lower packed layers of soil, etc. They will penetrate through hard clay areas, the worst of hardpans. I did some of the first home garden work with these unique chemicals.

Other additives:

We have dolomite, rice hulls, ground corn cobs, and all kinds of specially bagged materials for various uses from the shade garden to the vegetable garden. They definitely have their place. Some of the items may be hard to find as dolomite; some are related to given areas, wherever you live. Some are limited to an area because of scarcity or even weight. Such items as "Sponge-rock" may be somewhat limited as to cost. Some are very valuable for special things such as making cuttings, in house plant mixes, etc. One good illustration is the use of vermiculite. Used properly, with some knowledge of its use, it's fine; but used any old way, it holds so much moisture that it can rot out your seeds or cuttings. Again, its use as a soil conditioner is too costly. We, no doubt, will have future new materials to use—experiment with them for sure. Examine their life span, their cost, their ease in purchase.

Chapter Sixteen

MAKING COMPOST

The word compost comes from the Greek meaning to "join and work together". Briefly, that is exactly what it is. We are taking many materials and joining them together. You don't buy compost per se, you make it. It has been made for hundreds of years, and works from merely putting leaves in a pile, to specially constructed boxes. The "how" of composting making is as varied as recipes for cake making. As far as I am concerned, if the results are what you want—then follow your special method.

How compost making works:

There are basic things that make a compost work. The principles you can't dodge. They are "loud and clear", and as you vary them one way or the other, you either have a compost that works, or it doesn't.

1. *Moisture:* Compost is broken down by "rotting" which is pure action by bacteria and fungus digesting the materials. They work at their optimum best in damp materials—not sopping wet.
2. *Air:* There are two kinds of bacteria that make compost work, those that must have air, and those that do not. The reason for turning is getting air circulation to keep these two types working.
3. *Size of materials:* It stands to reason great big hunks of stuff will take longer for bacteria to break down than if they were finer. It is hard to break down limbs, pieces of wood, great stems of cabbage in a hurry. The finer the material, the faster they're going to break up into humus. That is one reason for "grinding" and "shredding" ahead of placing huge pieces into a pile.
4. *Heat:* As the compost works, heat is germinated. A compost will reach 165 degrees in its center. This heat kills any harmful insects, weed seeds, and may control some diseases. In

in keeping air going into the pile, turning or otherwise, the heat keeps going at a rapid, even pace. Too long a heat could also be damaging, killing certain bacteria.

How do we make it?

Keeping the factors of air and water before us, let's look at the various methods. None are wrong, I must stress that. Some just work faster than others. Some, in your backyard situation, might be better for you. Ken and Pat Kraft, in the great book "Growing Food the Natural Way", (Doubleday, 1973), say in their chapter on compost making—"If you want two different opinions, all you need is two experts on the same subject." When it comes to compost, this, to make a small pun, is true in spades. Everybody has his own idea of how to make compost and some methods are pretty involved. However, the differences are all variations on a theme. We're not going to say that most of the systems don't have merit, but we would like to make something clear . . . making compost is easy." They cover eight methods, including the Gordon Baker Lloyd method.

1. *The first compost making,* piled materials by the farmer: In the very old books we found farmers were piling manure, piling leaves, making wind rows of them, just covering the land with 6 inches of waste and letting it set for a year. Everyone knew they broke down, made the soil better, and stimulated plant growth.
2. *Better piles:* The English did a great deal in pioneering the use of everything grown. They began to talk about making piles with layers of this and that to help in decomposition. Home gardeners in large areas are still making piles of leaves. Some let them sit. Others will put in four inches of leaves, a dusting of earth, more leaves, etc. If these piles just sit, they will take a year or more to break down, and then not completely for the outer leaves will not break in this method any faster than they do in the forest.
3. *Digging holes:* Before we ever began to make boxes, we were digging holes in the ground pouring in leaves and stuff, cover-

ing with layers of earth to keep down flies. These holes often moved across a garden area. I did it as a boy . . . neighbors did it, and many reports right now can be found of such actions across the yard in soil preparation. It must be remembered that this is slow. It does work, we will admit, but it fore-cludes the use of compost in making potting mixtures, mulch-ing, etc. It will hold moisture faster than mere piles. If it is not dusted with earth, it will smell and draw flies.

5. *Odor:* If the compost box or container is smelling, and drawing flies, it is not working right. Layers of materials purifying and smelling is not making compost.

What to put into compost:

The forest floor:

Compost making reminds me of what happens on the forest floor. The fresh leaves are on the top layer you see as you walk over the soil. Just below, things are beginning to discolor. Then below that, you see leaves breaking up. Further down, they are almost gone. Then comes the invisible layer of fully rotted materials we call humus. These materials stay there year after year, getting moisture from the rain. If the season is dry, the layer sits. When it's moist, they begin to break down. Limbs fall, huge branches fall, trunks fall, and it takes years for them to finally turn into piles of material that work into the soil. Moisture, fungus, bacteria, insects, worms, the elements, break down the materials to return to the ground and eventually feed the plants.

Speed of leaf breaking:

Some materials will break down faster than others. Not entirely because of the size of material, but because of the material itself. Some leaves may have a heavier oil content and break slower. Examples would be walnut leaves, eucalyptus, acacia, and so on. It's not that the leaves are harmful—it's that they break slowly. Other materials such as lawn grasses, fresh waste material from kitchen, and average shrub and tree leaves, break quickly. Twigs and small branches break down very slowly.

What not to use:

There are some definite things not to use, period. It's not the question of them being slow, it's the question to avoid them entirely.

1. *Kitchen wastes:* Vegetable tops, fresh parings, can be used. But at no time the use of table wastes, wet garbage as it might be called. In some areas it is forbidden by law to bury heat wastes. The tendency in a compost pile is that some materials will build problems that will cause trouble.
2. *Diseased leaves:* Rusted material, etc. are carried in the soil. Keep them out of compost making. As far as I am concerned, I avoid anything that is diseased, mildewed, blighted, or what not. It will require some earth filling as the holes begin to sag in the breaking down process.
3. *The six sided boxes:* The next method to arrive was to make a big box with bottom and lid, an "all covered type". It broke down materials all right—smelly and horrible. It did hold moisture, sometimes too much. It did make a compact situation which was easy to dig out and handle. We have long since found that this tends to create terrific odor and draws flies—and although it works over a period of 6 months to a year, it certainly cannot be called the best.
4. *Four sided boxes:* As compost makers began to work with compost, it was found that if the bottom and top were left open, compost broke down faster. In this type of box, perhaps it could be made in under a year. Worms came up from the soil and helped it break down. Water was easy to apply and also layers of good earth. As the box idea began to develop, removal of sides took place. Not only for getting out compost easier, but to turn it. It was found that the more you turned, the faster the compost worked. (Look back at the principles we talked about.)
5. *Improvements on the box:* The box idea helped to hold compost in place and made a neat part of any yard design. If air was important—bacteria growth—and water was important, maybe

the box could be improved upon. It was here then that slatted sides began to be developed to let in air. Holes were put into boxes to let air and water get into the center easier. Wire enclosures were being made from chicken wire to heavy fancier materials. These containers began to turn out compost from 6 weeks to 2 months, of course depending somewhat on what was used.

Additives:

No one knows when layers of soil were added or addition of manure took place. All that was found out was that compost broke down quicker. In some areas, lime was used to hasten breaking of tough materials and keep down bugs.

1. *Lime:* In the actual chemical discussion of it's use, it will break down fibers. In our Western area, high alkaline spots, we had better forget its use for it does raise the pH. I leave this to your judgement. If you ask about gypsum, it does not do the same thing. Ashes are used but I would avoid them for the same alkaline reason. If a fly problem was very apparent, you could dust layers with ashes and then a layer of earth.
2. *Earth:* The use of earth is threefold. First a light layer, enough to cover the layers of leaves, etc. so they cannot be seen, keeps down fly problems. Eggs are not laid in the earth. Secondly, earth does keep down foul odors. Lastly, good earth is a bacterial inductor. It contains bacteria and when added to the pile, the pile breaks down quicker. It could be added that there are factors in using earth to allow for less packing and thereby getting in more air and water circulation.
3. *Use of manure:* Its value was to hasten decomposition. Added bacteria did this along with fungus materials, etc. No doubt, depending on type used, manures added some food value. I have many times hastened decomposition with chicken or rabbit manure. Because the "bagged manures" have salts, little food value, and certainly little bulk, I would avoid their use. Instead I would use materials like Nitrohumus (previously discussed). Here is a good bacterial additive and all you need

over your earth layers is a half inch of the material. It would be the best additive I could use.

4. *Compost makers:* They are available from little packets from England, to five pound bags on the West Coast. If you cannot follow the basic methods of making compost, they may be of some value. I am not one to knock any product. But I will have to admit I cannot see any difference with them or without them if you follow what compost needs to make it work. Speaking of additives, I will never forget the bottles of bacteria a doctor and I put up to make compost work better. No question they were loaded with bacteria and did replace manure. But from that point on—forget it! It may be that in the future we will have a real bacterial additive.
5. *Plant foods, liquid or dry:* There is plenty of proof that the use of dry and liquid materials will hasten breaking down of tissue. The dry materials can be dusted over every other layer. Use compost plant food, please, so you accomplish adding some food value. Using liquids is very simple. Add as you might water to keep the compost damp.

Summary:

You can make compost with air, water, and nothing else. Bacteria will do the job. Materials you add might hasten the breaking down quicker. The resultant end might be compost that is a bit richer.

The actual box:

You should take a look at the 8 methods of compost making in the Ken and Pat Kraft book "Growing Food the Natural Way". You should also read "Let an Earthworm Be Your Garbageman". Study the pictures in "Basic Gardening Illustrated" by Sunset, 1975 edition, and other materials for sure. I think "both sides" of any such popular subject need studying. Taking this all into consideration, during the war years I developed a box for compost making, Wood was scarce, so were all the additives. It has been written up many times, for which I say thanks, and I am going to give it here in detail. Again, may I add loud and clear—this is not the only way for sure. Someone may argue and say they have one better. Great! . . . but under many adverse and backyard garden conditions, I know this one works.

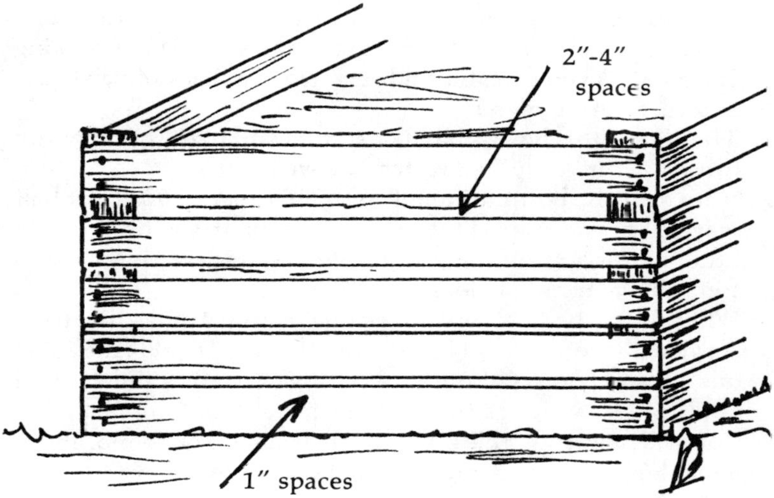

GL's box — portable

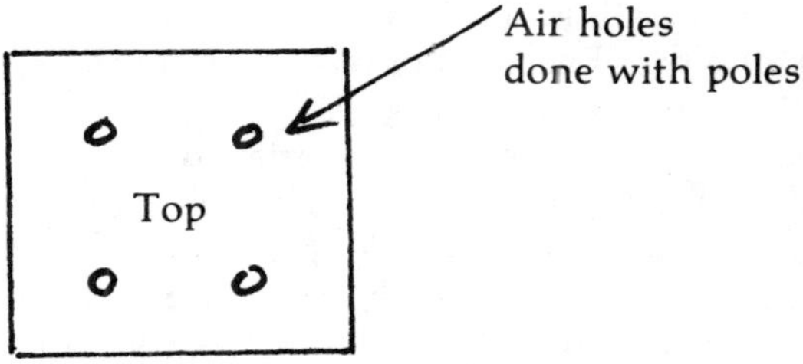

1. We are going to follow the theory that makes compost work—air and water.

2. The size of our 4-sided box can vary. I find one about 40 inches high, 40 inches square is about right.
3. To allow air to circulate we are going to make it out of slats. They can be all the same width—the trick is the spacing apart. Slats can be materials that are half to one inch in thickness and 3 to 4 inches wide. My first box was made out of packing crate slats. (You can still find such wood in back of hardware and furniture stores!)
4. The idea of this box is that it does not want to be so heavy you cannot lift it easily—so watch the weight!)
5. Slats near the bottom should be an inch apart, and about half way up get wider apart till near the top about 3 inches wide. In other words, a gradual widening of the slats between base and top. This is for all 4 sides.
6. When I first began these boxes I nailed the 4 sides together. As years have gone by I have wired them so I could take a side out . . . and friends of mine have hinged corners and done all kinds of fancy things to suit their problems.
7. You will need four poles, 1 to 1½ inch square or round, about 5 feet long. These go into a pile to let in air as per instructions following.

Making the GBL box work:

The box will be layered, 3 to 4 inches deep of the product you are composting. Then a layer of earth sufficient to hide the materials in. Every other layer you can add your activator. For a 40 inch square box, a large fruit juice can or half gallon, of Nitro-humus spread over the material followed by earth. After you have the first layer built, about 12 to 18 inches out from each corner you will use your long 5 foot stakes. To start with, push them into the soil just a bit so they are anchored. As the layers are built up you wiggle the stakes back and forth to make air holes. Leave them right in the compost. Wiggle frequently. In these air holes you can also pour your water. Keep pile evenly damp, not sopping wet.

When the pile reaches the top, let it set another week or so. No one can average out the time it takes to fill a box. But after filling, you can take it down, remove a side for 4 to 6 weeks, and follow these suggestions:

1. The best idea is to have two boxes working. So when one is done you are ready to start on the other—with leftover materials that are not quite rotted.
2. Lift up the entire box and break down the pile. To make the best compost, you will sift through ½-1 inch mesh or even hardware cloth. This fine material, you store or use.
3. The coarse materials go right back into the next box—or same box if you wish. They will rot down further. They also act as activators.

Use of Compost

The primary purpose of course, is soil conditioning. The finer the particles the better it gets into the soil. Hence, some sifting. Coarse products can be used in mulching. The very fine siftings can be used in seeding and pot work. If you intend to use it in seeding, and pots, it will pay to kill any harmful bacteria that can come from soil in the mix. This is easy to do by baking your mix in an oven at 450° for an hour. It will smell, but it works. You can also do a fair job with 1 tbsp. of Chlorox to a gallon of water, poured through your flat or mmaterials you use for seeding. Do the same for any potting mix for indoor use. For outside work, forget it!

A shovel blade must go in at right angles. It is then easy to pull back and break the soil.

PART FOUR

HOW TO DO IT

Chapter Seventeen

HOW TO DO THE ACTUAL SOIL CONDITIONING OR IMPROVEMENT

This discussion we will do in two parts. The first part will be the conditioning by hand, which most of us will do. The second will be conditioning by mechanical means—the power tiller.

Hand Digging:

For as long as any home gardener, and even beginning farmers can remember, we took our digging tool into the soil area, pushed it straight in, and turned the ground over. We frequently turned in nothing but weeds. As the years progressed, we added manure and other materials, and even still, we put on our two inches of material and turned in in and expected wonders! Why don't the materials work? Let's take a good look at simple digging.

1. Following instructions on the bag, we put out our stuff. With our digging tool, we dig it in. Where does it go? To the bottom of the hole. Does it mix with all the little soil particles? No!
2. Remember old time gardeners who dug the ground once and a month later did it again, and a month after that, once more. Some began in the fall and dug over and over all winter til spring came for planting. Why did this work? Materials were churned over and over and at least mixed from the top to the bottom.

The Top of the Dug Area:

In general digging, the top three to four inches gets none of the humus-forming materials. Therefore, the soil crusts quickly. It's hard to get seeds up. Moisture retention is poor. I might even say that if you do nothing but dig in things, about 80 percent of the dug-in materials are of no value to the first half of the life of any plant. If we practice some type of "churning", then the

REDWOOD
SOIL CONDITIONER
fortified with .5%N
organic nitrogen
LINDAUER & CO.

materials we add will be of some value in seeding and small plant root formation. As many an educator says, give me a child to the age of seven, and you will never have to worry about the character of the child. Give me a plant at the first one-third of its growing time, and I will give you a plant you don't have to worry about later!

The Lower Part of the Dug Area:

The deeper the soil conditioning, the better the root system. The deeper the roots, the better the plant. Throughout the length of the spade or fork depth, we need our soil conditioner to work—organic matter and so on, but what about that layer beyond the spade depth?

What About Hardpan?

This is the area at the bottom of the normal shovel or fork depth, roughly eight to twelve inches in. It can come from fill soil; it can come from natural subsurface conditions; it can come from soil layers. But whatever it comes from, it will hold back root growth. If you don't know a cross section of your root soil, now is a good time to do it. Dig in a foot or eighteen inches. Pour in water, half up the hole. If the water hasn't gone out in 45 minutes, then you had better take a hard look at what's below. Oftentimes, in orchard planting, this may mean tremendous large holes done with soil drilling equipment and even blasting to get holes large enough, where roots can expand in normal growth and, most of all, good drainage. I mention this to show its importance.

How to Help Our Hardpan:

If I was going to say that the ultimate in soil conditioning was to dig two feet deep, regardless of any kind of soil, we would all go right up the wall! When I first began to teach soil work, I did just that. The more severe I got on it, the fewer I found who would do it, and I gues the older I get, the more I can understand why. It's hard work! On the other hand, two-foot work is permanent! The English have done it for years, and back in their old books, they talked about "double trenching", "bastard trenching", and so on, with great drawings and details as to how to do it. That was then, but look at the pictures in Sunset's BASIC GARDENING ILLUSTRATED, 1975 edition. We are doing it again! In heavy soils, this is most valuable. Perhaps the great success in raising beds is just this. Soil is mixed correctly. Drainage is perfect. Bacterial action is high. We have the ideal soil!

If You Can't Double Dig — Do At Least This

Minimum amount to spend on soil work (little money to spend):

1. Always have your soil damp when you dig, spade, shovel or fork.
2. Dig soil up and rake down, breaking up largest clots and taking out any heavy debris. No, you do not have to take out the pebbles—just the larger rocks that are in the road.
3. Add one to two inches of your organic matter. Churn this in with a grading rake (rake with straight teeth that bite in), a three or four-pronged cultivator, or even a hoe. Mix and churn this material into the top two to four inches of soil. Be sure each particle is mixed against each particle of soil, light or heavy.
4. Level and get ready for planting.
5. At least the materials you have mixed in will help in getting the lawn seeds to start, the small flower and vegetable plants and even the over-all shrub border, etc.

Maximum soil work (more money to spend on the length of the shovel, spade or fork):

1. Have the soil damp.
2. Apply one to two inches of your materials and dig them in the full length of your digging tool.
3. Rake down clods as before.
4. Apply another same of material and churn this into the top two to four inches, as before.
5. Level for planting.
6. What we have here is a good cross section of fairly mixed soil

so that both seeds and seedlings can grow and as the roots go deeper, they will have some help.

Tools to Use:

Before we go any further, let's talk about tools to use in relation to our soil types.

1. *For heavy soil:* The best digging tool is a long-handled, straight fork. The best fork has square tines, not flat. In either instance, you want a fork with long, straight tines. It will work the easiest into adobe or clay-type soils. If you can push in a shovel or spade, fine.
2. *For light soil:* (easy to work soils) The best tool is the old fashioned spade. This should be long and straight and handled for sure. Next best for use would be an irrigation shovel, which is almost straight with a handle. The shoveling shovel, which most of us have, is made just for that—lifting and shoveling earth and materials.

Short versus long handles:

The long handled tools give you the best fulcrums and leverage. Short handled tools require lots of back bending. They are great to carry in the car for fire and other emergencies. Don't use them in the garden!

Handles:

Actually, the more you pay for your digging tool, the better the steel and the better the handle. Do not buy painted wood handles. Look for long grains. Check on handle replacement, in case. When you buy your digging tool, buy it for years to come and treat it accordingly!

How to Dig

This sounds silly, right? Sure, we all can dig a hole. We have all dug over our ground. So how's your back? Hurt more after an hour? Then you don't know how to dig! The best way to show you how is by the pictures I have for you.

Three typical digging tools. Gordon holds irrigation shovel, the best tool for average digging. Next in the middle, a fork for heavy soils, and Judy holds a spade that is great for edging but hard on digging.

Mechanical Digging:

By mechanical digging, we mean the use of a tiller of some type to go over the areas thoroughly with a churning, mixing operation. We do this in lawn making. Our flower or vegetable area preparation, and we must admit it does save time and does a thorough job. We can do this in two ways. We rent equipment; a half day generally does it, or if we have a small orchard or larger vegetable areas, we will own one that we use off and on throughout the year. It stands to reason that if we are professional in our field of landscaping, etc., we will have one as part of our equipment.

Renting:

You will have to take what's on hand. Explain the size of the area you have to do and get some information as to how deep it will till—and how it works!

Purchasing:

From long experience, I suggest this:

1. Buy the tiller to suit your job. How extensively will you use it? This will depend on the horsepower you buy.
2. Buy something that is not new on the market, but has a good background and service availability. Please buy only from a reliable dealer, to be assured of future help.
3. Be sure of the guarantee, length of time and what does it cover.
4. And obviously, fit the tiller for your own area and the job you expect to do.

To Dig or Not to Dig:

This is going to come up for sure. We have several new books that say absolutely not! Perhaps so. Maybe we don't need the bottom soil at the top. We may not want the lower soils in our seed bed. We don't need weeds hidden from lack of air and water. This I admit, but what happens on several yards is not the way it may happen on yours and mine. I do know that the better we prepare our soil, and the deeper we do it, the less likely we will

If you use a shoveling shovel, you have to bend forward so the blade goes in at right angles.

have to go through the horrible job of intensive soil preparation we originally might have to do. We can keep up sources of organic matter with digging. So, again, give it thorough thinking, and don't jump to the first conclusion. Yes, we will be digging and doing it with some reason and thought.

Chapter Eighteen

CULTIVATION AND MULCHING

Cultivation

Why we cultivate:

The word cultivation began to be used way back in Biblical times. Old books talked about it, even a volume or two written for farmers as to how to do it. It was the consensus of opinions that cultivation did several things. For sure, it kept down weeds. Then, in very dry, rainless areas, cultivation made a dust mulch that held what little water was available from rains. When it rained, everyone was happy. Right after the rain, you rushed out to hold it in the soil by light cultivation. This is beginning to die out for soils that have been cultivated and cultivated have top areas that are easily blown away. Backyard gardeners cultivated to keep down weeds and to break the crust of the soil to hold moisture and to let in air and water. Any soil that had a crust kept back water absorption.

Crusted soil and cultivators:

Since soil must have, by content, 25 percent air and at least 25 percent moisture, anything holding back absorption of either requires a remedy. When a soil has a crust, air bounces off and it cannot get in. Hence, the tickling of the top of the soil to let in air. Water is not readily absorbed into a soil if there is a crust, so we stirred the soil to let in water. To help in this, we had little, small-type plow cultivators, three, four or up to six prongs. These we pulled through the soil very much like little plows. Then came two, three and four straight-pronged cultivators to stir the soil lightly. These are still in use and have a great place in stirring in soil amendments. Bringing this into focus, I am showing some of the "gadgets" from my very broad collection of garden tools. (Comments are my own . . .)

Mulching

Just where mulching began, I cannot pin down to an exact date, but when it started, it was found that mulching did the same thing as cultivation, only better. By mulching, we mean the addition of organic matter to the top of the soil in depth of one to four inches depending on climate and some factors we will discuss in detail. I want to remind you that mulching indeed enters *into the continuation* of soil conditioning.

Here is what "you are hearing now on mulching"— the current mulch discussion:

1. Right off the top of what a mulch does, comes the factor of moisture holding. This is what everyone is told loud and clear. Mulch your shade plants to hold water. In our current years of vegetable gardening, mulching has come to the foreground more than ever. So we hear, mulch around your plants to hold moisture. It makes the plants grow better!
2. As we began to mulch, we found there were less weeds. "Mulch to stop weed growth!", was another cry.
3. In the last ten years, we have been told over and over by organic enthusiasts that you must mulch, even "grow in a mulch" and the deeper the better. Here is an "ideal situation for plants to grow", etc., etc. . . .
4. The final arguments for mulching include less pests, a "more natural condition for growing plants", plus some wild things that we won't need to discuss.

What does a mulch really do?

Let's take a look at what happens in nature, which everyone likes to quote. It is true that the accumulation of leaves, as we have already discussed, holds moisture, prevents packing and lets in air, and, finally, breaks down to form a humus and the resultant food to the roots of the trees, shrubs and flowers. With the help of rain, snow, wind, fire, and the other many side effects in nature that we don't consider, this natural condition manages to work to keep our forest areas, the meadows, the areas we all

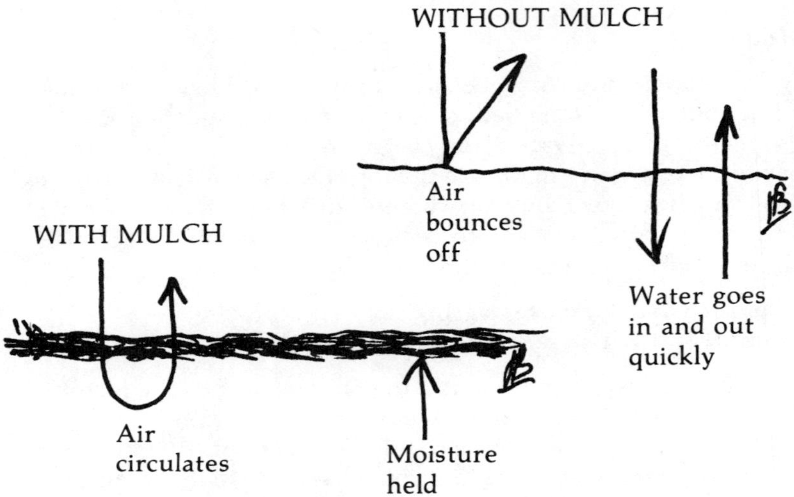

love to visit, in the conditions we take for granted. When we have fires that wipe out our undergrowth and mulches, we are in trouble. Then, when we get too much rain that washes them off, we are in trouble. With no rain, we have less growth, and resultant problems. So, with this in mind, let's look at mulching, under even more ideal conditions than in nature. For this we can be very thankful. As we discuss mulches, let's "use our heads", looking at mulching as we find it in ideal conditions in nature.

1. *A mulch does hold moisture:*

A good example of this is to kick up the layer of leaves under the forest trees. Another example is the moisture under the board, a piece of burlap or the stepping stones. These are types of "mulches". Mulches do hold moisture in the ratio of how they are made. Some coarse mulches, such as straw, three inches deep, may hold little moisture. An inch of redwood materials, barks two inches deep, compost one to two inches, may hold moisture nicely. This is going to be largely up to you and your own conditions as to which may work the best. A mulch must allow water to go through it, during rain or your own watering.

Some mulches, then, defeat the purpose of a good mulch. Some mulches are so coarse that water is not held at all. They are for decoration. Maybe we can say, the finer the particles of the mulch, the better the moisture holding quality and, perhaps, the less we have to use.

2. *Mulches do control weeds:*

Seeds can sprout under a mulch, but they cannot grow up through it normally. In other words, the mulch inhibits growth. One factor of leaf drop in nature is weed control. If weeds grew in mass under all trees and shrubs, the poor plants would be fighting for moisture. In the new vegetable garden mulches are a must for they help to control the weeds in the newly turned soil.

3. *Mulches help air circulate:*

One reason we break the crust in cultivation is to let air in. A crusted soil repels air; the air bounces off. Mulches keep the top soil pores open and air circulates in the normal method—just as in nature. Remember, air is needed by roots as well as by bacteria and certain fungus.

4. *Mulches help in temperature control:*

There has not been as much research on this as I would like to have seen, but we know from years of use that we mulch to protect damage on the bud area of roses and other plants, and in heavy snow and cold areas. In the yard, mulches do much the same thing in temperature control. A one to two inch mulch will keep out cold. I did not say heavy snow cold! A mulch one to two inches will keep out heat. This is most important in all gardens in the West in summer. When a soil gets hot, roots burn. They may not always shrivel up and burn as with a blow torch, but the tender ends can burn. When this happens, the plant is in a state of shock, and if it keeps up, it can be killed. Bacteria ceases to function in hot soil. Plant food is not well absorbed in soils that reach high temperatures. We say that the more even the temperature in a soil, the better the root growth. One reason we use bottom heat at each end of the year to get up seeds is that even temperature. How important this is in the shade garden! Look at nature and the deep mulches. How well they work in the hot weather, so you may mulch to hold moisture or cut down weeds, but temperature control is almost as important. Mulches will keep roots warmer in winter. Here, you must use some discretion in its depth for you could keep the rain from going into roots.

5. *Mulches do add organic matter into the soil:*

We go back to nature again. Noone is "out there" digging. The mulch—the litter on the forest floor does something. It goes back into the soil. I am not ready to agree that in soil conditioning, all we have to do is to mulch and let it work. Somewhere along the line, the promoters of this idea forget that time is an element. We have to do some very intensive work to get the conditioning started that has been neglected in soil for a hundred years. However, after you have done a good job of thorough soil conditioning, it is remarkable how mulches work in adding further organic matter with no soil turning. A good example would be in the shade garden. We cannot be digging around these sensitive, shallow-rooted plants. Mulches do the job of adding organic matter to the soil. We should not be digging in the shrub border, the foundation planting, for mulches will add the organic matter to the soil. We should not be digging in the shrub border, the foundation planting, for mulches will add the organic matter needed. In pot and container grown plants, we have found the addition yearly of an inch of mulching materials or the addition of the mix does wonders without having to take the plant out and repair the soil. How important mulching is in the area of ground cover planting. What do you do with the lawn to get organic matter back into the soil? We cannot just leave the leaves for we would have all kinds of disease problems, so what do we do to get into this very complex, tight root system and organic matter? We top dress, as it is called. It's the same as mulching. We do it in light amounts, but where does the dressing go? Right down into the root area. We, at home, may do it only once a year, but for real great greens on the golf course, you ought to see their program!

6. *Mulches may add food.*

It stands to reason when the mulch breaks down into humus, some food value is obtained. Again, this has a lot to do with what you are using for a mulch. As in nature, this is a good last reason for mulching. I point out, we do not depend, in the backyard garden, on the mulches for our plant feeding. Perhaps, after ten years in a given bed, we reach the ultimate, and the mulch does the job, but with water factors and all that goes with modern civilization, let's use mulches for the main purpose as we have discussed.

7. *Mulches in pest control.*

There is no question that mulches can attract certain night-feeding creatures. Layers of leaves will certainly attract earwigs, snails, and slugs. The properly worked mulch, with fine-textured materials, does not attract insects. I am not interested in some of the "way out" mulch materials being suggested. I have seen some pretty bad insect damage done with layers of newspapers and even plastic mulches in backyard gardens. Experiment on your own, but don't spread your results as conclusive theory for everyone to use because it may not work. We do put out damp burlap and even boards to attract earwigs for easy hand control. If the resultant breakdown of the mulch into humus and then into plant food helps build up resistance to the plant to insects and disease—great! However, let's not say that mulches help in the pest control.

Summary:

We can say that materials mulched grow better than those that are not mulched. They may even grow twice as tall and fruit larger. Roots grow thicker, and the plant is by far much better. The reasons we have given, I think, are obvious. Now, let's use our mulches with some sense and reason.

How To Mulch

I always enjoy hearing the mulching overenthusiasts telling everyone in ear shot, "Just put on four to six inches and you have the garden made for sure!" Why not do it right? Who says right is right? Well, then let's say common sense. The guy that simply spreads his mulch around and has such good luck may be mulching on soil that is already nicely open. In other words, the soil may have been previously dug and only slightly settled. He may be mulching on soils that once had a mulch. A mulch simply does not work as it should when but over bare, undug, hard-packed soil, sand or adobe. It doesn't work that way in nature, and it won't in the backyard. If we kept doing it for two years, it might begin to work, but we want our mulches to begin to work now. So try these ideas:

1. In the newly prepared garden soils, vegetable rows, flower beds, etc., add one inch of materials and gently tickle them into the soil about an inch.
2. Now add another inch of the same material. This gives you about a two inch depth of mulch. If it seems to be a bit shallow, then add a bit more so you have two inches of apparent mulching materials.

This procedure is joining the mulches to the soil surface. If you don't do this, mulches can slide off, float off, even blow off and you have lost what you are after. I might add that in a strong wing mulches can blow away even though you use this procedure.

Mulching in High Windy Areas

This is an exception to the rule, but if it happens, you put over your regular mulch some materials that hold it down. This would be the use of bark materials. Look for half inch to one inch "pebbled bark" to hold the mulch down. I see chapters written on this subject and wonder where the author lives. Outside of some desert areas and perhaps a few windy times in any yard, this is certainly a negative approach to mulching. They don't blow away that often!

When natural mulch is not available you add your own. Material used is rotted redwood sawdust.

What to use

You will want fine textured materials that do the best job. We start with those that last the longest:

1. Rotted, leeched redwood sawdust—1 to 2 inches;
2. Peat moss—1 to 2 inches;
3. Fairly fine compost—1 to 2 inches;
4. High quality leaf mold—1 to 2 inches.

Other materials from rice hulls on depend on your area and your experiments. I have confined myself to those materials widely available.

Mulch Placement

Vegetables: Place mulch in irrigation ditches, in beds (melons, etc.). Mulch over entire area.

Berries, grapes: Same as for vegetables.

Flowers: Mulch in watering area. Mulch in between flowers in beds.

Shrubs: Mulch over entire root area. This is very important in the shade garden.

Trees: Young trees cover entire area. As trees get from three to more years in age, mulches had better be kept away from the trunk. Mulching materials against a trunk can bring rot, diseases, etc. Keep away six inches to a foot. This is very important if you are a deep mulcher.

Containers: Mulch at least an inch; same for pots, etc.

Lawns: Put on one-half to one inch fine material at least in spring and winter. Three times a year would be even better.

General instructions:

Your mulches work even better if you will mix in one half Nitrohumus. You are adding organic matter plus life.

Chapter Nineteen

ROCK MULCHING AND SOIL CONDITIONING

What is it:

If you have never heard of rock mulching, don't let it worry you. It is very evident many haven't. There is nothing new about it for it was practiced many years ago—several hundred years ago, in the steep sloped area of the Mediterranean, Italy and such places. I first heard about it in the years right after the war when Herbert White began to write about it and show its use. When he passed away, the good word sort of stopped. I did a yearly discussion on it for many years, and it was picked up with some short notations in Ken and Pat Kraft's book, "Growing Food the Natural Way". There are some pictures with comments in the 1975 edition of Sunset's "Basic Gardening Illus." I have been told there is some further material on it by some of the organic enthusiasts here and there. As far as I know, I have brought it down to some simple procedures and facts that are not found elsewhere. Rock mulching is just that . . . mulching with rocks. It's a job, but worth it over and over. It does have some pros and cons, chiefly in the range of work and getting the rocks.

How rocks on top of the soil work:

Instead of using an organic layer per se, we are going to use materials that are completely different. A rock, in its thickness, would prevent heat or cold from going through it easily. A rock on top of the soil holds moisture underneath its base. Remember, when you wanted to find unusual bugs? You looked under a rock. The rock, then, is going to repel excessive temperatures. This is important in any root growth, as we have studied, and it is important in keeping up high bacterial activity. Even temperature in root areas does wonders. It is important to consider thickness of what we use. Thin, two-inch slab-type rock would let heat go through easily and might not keep out cold.

Some types of rock that are porous would defeat what we are after. Just laying rocks on the surface of the soil would keep out weeds, would help in holding moisture, would help in keeping temperatures more even, but what hs this to do with soil conditioning? Here is where few have discovered what this can really do.

How to condition soil by rock mulching:

The best way to work up to what this type of mulching can do is to tell you, step by step, how to do it. Then I can explain the end results. Let's take a fruit tree that has done practically nothing over the years—badly neglected and just about ready for the axe.

1. From about eighteen inches out from the trunk of the tree, hollow out an area roughly six inches deep beyond the drip edge of the tree—end of branches—at least four feet. Don't dig so deep that you are digging into masses of roots. You have to get down into the soil at least four inches, roots or not. There is no value going out beyond six inches. The soil you remove you can store, use in compost making, give to your neighbor, or maybe use in raised beds or a fill in your garden—or perhaps a planting mound that might help to keep down on noise out front.
2. In this dug-out area, you can bring in all your leaves, dead or fresh, weeds, straw, grasses, etc. Use anything except diseased materials. You are going to use materials just like in compost making, except you do not have to chew them up for this use. Try to avoid twigs and branches, for you are not making any contribution to humus forming forming materials in this process. Here is a good place to use barnyard materials with straw and bedding materials. If you have dug six inches, put in a six inch layer; if you have dug four inches, put in a four inch layer of materials. Add 1 inch of sludge, such as Nitrohumus. In other words, you have filled in your dug area. This is the best way to do this operation, if you have never done it before. As you get experienced, you can decide on how much to add so that the rocks give a better esthetic

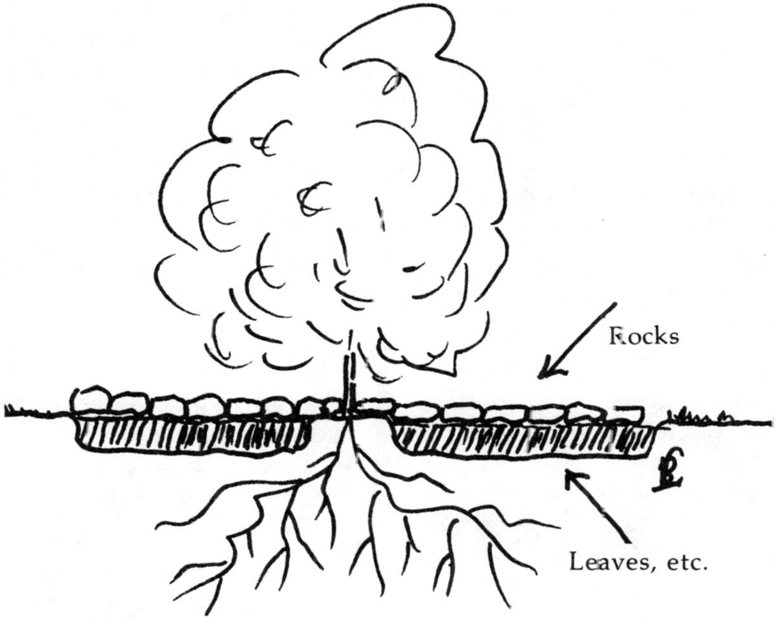

Cross section —
Note dug area goes beyond drip edge.

appearance.

3. Now, on top of this "mess", you are going to put your rocks. What rocks should you use? This is very important and here, perhaps, comes the limitation to rock mulching.
   a. Rocks want to be of very solid materials. Creek rock is best, made of granite. Porous rocks defeat what we are after.
   b. The ideal rock would be four inches in thickness and six to eight inches wide. Round rocks can work, but they do not layer too well.
   c. Will bricks work? Yes, but to a certain extent they defeat moisture and heat control.
   d. What about slab rock? They will work, but again, heat and cold will go through them. They do hold moisture.

4. What about design? I feel that just laying the rocks as close as you can over the area I have suggested will accomplish what you are after. If you get into designs, then you have several problems, such as how long will the design last; and the design will defeat what we are after in soil conditioning. We are not using rock for artistic purposes in this program. Maybe, after we do our conditioning, we can consider the looks, and some artistic rock mulching can be great!

How you make the rock mulching work:

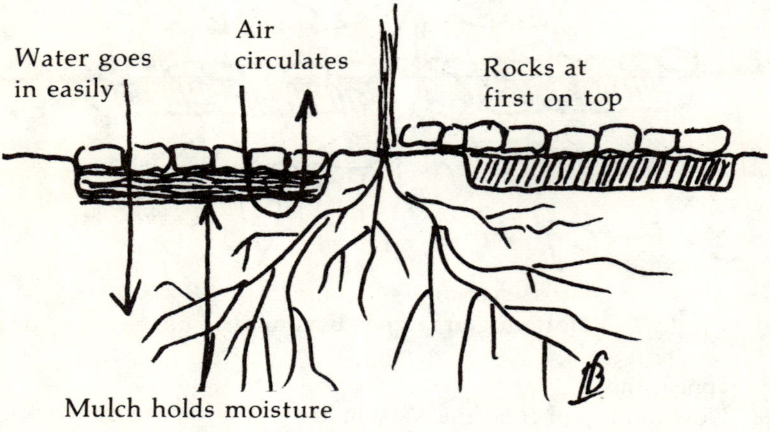

How mulch works . . .
As rocks settle, re-add mulch.

The rocks are riding on top of this layer of organic matter, which I called, with tongue in cheek, a "mess". There are things that you must do to make the soil conditioning take place.

1. Over the rock area, soak with your hose. To find out how much watering you have to do, depending again on weather and the various times of the year you may start this program, raise a rock and see what the moisture is like in the layer of

organic materials. Just keep them damp—not sopping wet, but damp.

2. Your feeding program is normal as always. If you are using a liquid food, pour it over the rocks and wash it in. The same can be done with dry food. Hard-to-dissolve foods will mean you have to raise the rocks and put it under them as you feel needed.

This is what begins to happen:

In areas where it is impossible to dig because of the soil type and the roots filling in the spade depth area, this method of soil conditioning is hard to believe. Here is what happens:

1. The organic matter begins to rot down. Earthworms come up into the area to hasten the decomposition. In fact, the earthworm activity is doubled very quickly. Rotting takes place from even moisture and temperature conditions.
2. The rocks, riding on top of the soil, begin to settle. As they begin to settle, you have a signal to add more materials. At first, the settling might take place in a month. It could be in two weeks, as I have observed many times.
3. Raise the rocks and add more of the same type materials. This time, you may not add quite as much material.
4. Eventually, the rocks do not drop as often and perhaps adding materials twice a year will do it.

Here is what has happened:

What you notice is that the materials you put under the rocks have begun to rot. In the first several applications, the materials seem to disappear quickly. This is because the soil needed organic matter, and it has literally burned it up. The action took place by being under an ideal mulch plus the tremendous work by worms going back and forth in the soil. The reason for the creek type of heavy rock was to give the ideal condition for the rotting to take place. You had even temperature, even moisture, and bacterial action started in a hurry. The slight cracks between the rocks not only let in moisture but also air. Finally, when the soil reaches a saturation of organic matter, it slows down just like with mulches on the soil in any garden.

What happens around the tree:

A somewhat ideal situation has taken place. Moisture is being held. Life is getting into the root area along with plant food. I have seen extremely sick trees begin to recover and take on new life. No longer is digging needed. The rocks accomplish the entire job.

Objections to mulching:

1. The original labor!
2. The nuisance of having to take up rocks to put in mulch.
3. Leaves that accumulate on top of the rocks.

The first is obvious. Sure, it's work, but the time it saves in the long run is worth it. You have to mulch anyhow, and often the mulches don't work around trees. This one does. The leaves that accumulate are easy to get rid of in the compost pile. The rock area doesn't have to be swept. It will look neat. The point of the whole thing is to get fruit results, and you do!

Other uses:

You can mulch areas between rows of vegetables. You can do the same with cane berries. It can be done in the rose garden and even in the shade garden. The use is up to you, as far as you want to go in experimenting. Your climate, your soil, your slope, your own problem controls how far you want to go. The few fruit trees you have, this is an asset. Is the soil conditioned? And how!

PART FIVE

KEEPING SOILS IN PRIME CONDITION

Chapter Twenty

SOIL NUTRIENTS — PLANT FEEDING

We are interested primarily in this book on soil conditioning. But we do have to realize that part of any soil are the nutrients that make plants grow. I am not going to go through every element and what it does. You can find that out in the reading of any general garden book. I want to focus on the easiest way to do it—for I believe this will save a lot on the complexity that seems to surround feeding.

How do plants eat in nature:

The leaves fall; the leaves break down into humus. Humus brings out the bacteria that attack the minerals in the soil and breaks them down. With moisture the food is taken up into the plant. This is it very simply. We know that light and temperatures are other factors. Does anyone go around feeding shrubs, grasses, wild flowers and trees in nature? No . . . we say they are under natural conditions and work "with nature".

The back yard soil:

It is pretty bad, normally. It lacks organic matter. It lacks plant food nutrients. In good soil conditioning we bring the soils back to a more normal condition . . . and this should include nutrition. Soil tests, taken professionally in a laboratory may show your soil high or low in certain nutrients, but unfortunately that has nothing to do with availability. In your backyard soils, we are able to get higher results in leaf, flower, food production, because are able to do a better job in concentrated gardening —which means better soil conditioning. This type of soil conditioning is very hard to do over large farm acreage.

What do plants need in nutrients?

Depending upon who you talk to, we recognize 12 to 17 elements. There are three very basic ones, nitrogen, phosphorous and potash. We call them basic because they are the ones used the most, in the greatest amount, and are the ones commonly listed on the plant food package. However, the other elments, may be minor in amounts found in the soil, but are far from being in their value. Very little amounts of boron are needed per ton of soil, but leave it out and you're in dire problem.

How much food do our plants need?

I must have 30 volumes of long technical discussions on what formulas are needed for what crop. We can argue the best indoor plant food formula till Hawaii freezes over. We can argue lawn foods, vegetable foods, rose foods, and you name it. Where does it get us with our backyards and small areas? *No where.* Thinking the entire thing out as we should, we will have to admit a good maufacturer of plant foods, with their laboratories and research, surely can compound a food that will work for us all over our garden area. There can be some exceptions, but the "all around", all purpose plant food is all we need.

1. My work with formulas over the years, tend to prove that those in the lower figure range, 5-10-6, etc., accomplish what we want without the problem of burning.
2. We know for sure that frequent amounts of plant food in small amounts are better than infrequently given large doses.
3. If formulas contain organic foods, they last longer in the soil and do not burn so easily.
4. If the plant food has minor elements in it, we are just that much further ahead.

1. *What does the formula mean:*

The first number on the formula is the percentage of nitrogen. This is what makes the plant green—it's what makes it grow. Some lawn foods are very high in nitrogen to push up the grass. On the other hand, using those very high ratio formulas over a period of time finally brings an opposite result. They stop working. There may be some crops where extra nitrogen can be ued with no long-range harm. But do you, in your backyard, understand which, when and how? No. So we leave the judgement to our manufacturer. The second numeral is for phosphorous. This is what makes for root, fruit and bloom. In most soils it can be found, but it's availability is nearly nil. Many foods are high in phosphorous. Such as 4-12-4, etc. This last numeral is potash. Briefly, this is what builds general strength in the plant, and one of the most important growth factors and often overlooked. In looking over soils over wide areas, a manufacturer compounds what he thinks is the best average food for use on everything in the garden. Why do we have peanut food? Rose Food? Lawn Food? Having sat more than once with producers in formula discussions—it's no great secret. It's pure pressure from home gardeners, from salesmen, from all over, that "this and that" is needed because it's easier to sell. You go into your nursery and tell them you have a lot of roses. Would you buy an all purpose food or rose food? Maybe I am letting out secrets, but that's the way a lot of foods are formed and products get on the market. This is no reflection on our first line manufacturers who do their level best to produce what we need and want. Added to his basic formula we hope can be found minor elements.

2. *How often do we feed?*

First I want to point out that plant foods applied, liquid or dry, do not stay in the soil too long. They are washed out by moisture, broken down by other factors, and are absorbed up into the plants. Even the most organic types, eventually leave the soil. I can remember years ago we told folks to feed the lawn three times a year at the ratio of say 3 lbs. of dry food to the 100 square feet. Then we found out that if we said six times a year at the ratio of 1½ lbs. to the 100 square feet, plants did better. It was not the question of using more food, it was giving plants little amonts of plant food frequently. Isn't this like what happens in nature? Early gardening pioneers, using little amounts, wer often growers of fuschia, begonia, and african violet. For years, they were feeding little amounts of food, every week. At the writing of this book I was looking over hundreds of acres where several million shrubs were growing in gallon and larger containers, all being fed little amounts of plant food whenever they were watered. One of our largest camillia growers in the country has been feeding weekly for the past thirty years. One of our best carnation growers has been doing it for twenty-five years every time they watered. Why does it take so long for this all to leak down to us?

It is very simple. If you are told to use 1 tbsp. of plant food to the square foot under your rose bush, which is in bad shape, you use a handful to the square foot instead—you get into trouble quickly. It can burn. It can do damage. And we blame the plant food producer. This is the first reason it's been slow to change instructions.

As a whole we are miserable instruction followers. Little amounts of anything doesn't dawn on us too well. As a general rule we could take any plant food and cut it in half and use it twice as often. Here is how often on plant feeding:

1. Trees . . . at least four times a year. Fruit and shade too.
2. Shrubs . . . at least twice a year.
3. Lawns . . . four to six times a year.
4. Flowers . . . monthly.
5. Vegetables . . . monthly.
6. House plants . . . every two weeks to monthly.
7. Roses . . . monthly.
8. Heavy bloomers like begonias, fuschias . . . weekly to every two weeks.

3. *Placement of food:*

We want to put plant food over the root area. Food, as well as moisture, is absorbed only one place, the 100th inch tip end of the root. The greatest area of tips would be at drip edge and beyond. This is why we feed shrubs and trees in bands around the drip edge—edge of the branch spread—and even between rows to get to the greatest root area. We call this in vegetable feeding and flower feeding—side dressings. We are feeding in between the rows in the watering area.

4. *How to use:*

We say to apply dry food on damp soil and soak in. This is to prevent root burning. Liquids we can use without previous damping of soil. Dry food can be worked into the soil previous to planting or seeding. Liquid food can be used at time of planting and as needed thereafter. There is very definite trend to liquid feeding. Commercially it is used as watering takes place. In the home garden we have proportioners that work both through the hose and at hose end. We have a new one that measures out 25 gallons of solution from a very small container. Inside the house we measure in the usual way. Most of our indoor plant foods are either liquid or quickly dissolvable. Some house plant foods are in tablet form and are pushed into the soil. We have a new trend to pellet food that is called "time released" for both indoor and outdoor feeding. My personal feeling is that more research has to be done with them in backyard soil conditions.

Chapter Twenty-One

ABOUT WATERING

I don't feel we should do a detailed chapter on watering; perhaps we can cover that in another book sometime, but we must discuss watering in relation to our soil conditioning. Since 25 percent of good soil must b moisture, it is essential we condition soil to get water to the plant properly. Roots have to reach out and grow to get moisture. In poor soil conditions, we get into all kinds of problems. With too much moisture in poorly drained soils, roots rot and die. In such soils, bacteria quit. In soils where water runs through too fast, plants starve and die of thirst. Bacteria begin to go dormant. So, moisture and soil conditioning go hand in hand.

Let's take a look at basic soil types in relation to moisture movement:

1. *Heavy soils:* Water moves slowly and evaporates slowly. One inch of water will penetrate about 4 inches. Time wise, it will take about 8 hours to water 4 feet deep around the fruit trees, and water will evaporate in about 4 to 6 weeks.
2. *Light soils:* This is the other end of the spectrum. Here, water moves quickly. One inch of water here penetrates 12 inches. To water 4 feet deep, you can soak it thoroughly in about 3 hours, and water will evaporate out in 2 to 3 weeks.

In other words, we can see that water movement is controlled by soil types. We are not discussing unusual layers, unusual blockage of water. We are talking about a general full length depth of a given type of soil. If you have a foot of heavy soil, followed by a layer that is very different, all the watering in the world would not work till that impervious layer is broken.

Depth of watering:

If the root tips absorb the water, it behooves us to water to the depth of the tip absorption area. It is interesting to note that as we learn to water deeper, roots follow water depth, get into a cooler area, and plants as a result can stand much greater shocks of cold and heat. These just do better than shallow rooted plants. The bulk of our annuals, perennials, and vegetables want water at least 2 feet deep. Shrubs—2 to 3 feet; fruit trees—4 to 6 feet deep. Shade trees the same and even deeper because of their tremendous root system, six to fifteen feet deep. Lawns, 6 to 8

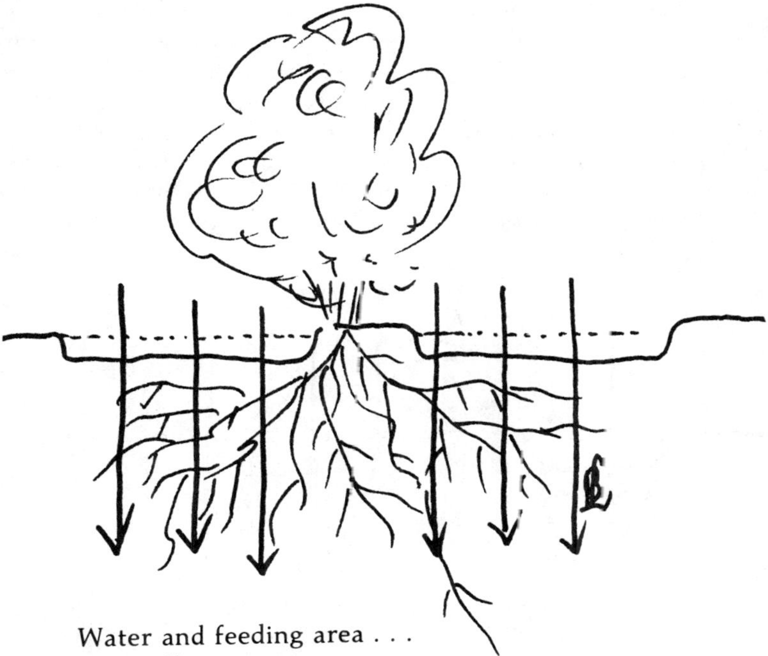

Water and feeding area . . .
cover all root area

inches to get deep roots. Seed beds, depth beyond the seeds. In other words, we cannot be watering shallowly. When we do this, we bring roots to the surface which causes all kinds of problems from adverse drying out, lack of plant food, absorbing, etc.

Spread of water:

This is as important as depth. Roots go beyond the drip edge or edge of the foliages. Have you noticed, in modern orchard watering, they run water in the middle of the tree row area? Notice placement of water in any vegetable area commercially grown. If we could say this is the ideal way to do it, it would be to cover the entire root area with water evenly placed. The exception would be right up against the trunk where there are no roots to absorb water. We can call this flooding. Since water falls heart shaped and does not spread too much left or right, it is important then that water falls over the entire root area. This begins to place a lot of questions on more ditches around shrubs and roses and even fruit trees. The water is limited in its placement. The plant survives rather than grows as it should.

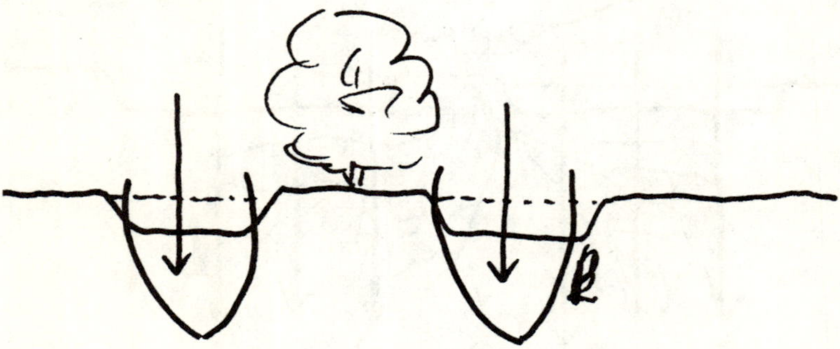

Water goes into soil heart shaped.

Those lovely large flowers and fruits and vegetables:

This may be hard for you to believe, but most of the lucious growth we like, smell and eat, comes from proper watering and not so much the feeding program. The difference between the florist flower and the one you grow is not variety or some secret ingredient. It's the great depth of watering in between those ridges. I cannot stress enough the importance of water study in relation to plant growth, and your soil conditioning controls this for sure.

How to do it:

With hundreds of watering tools to use, which do you use? Let's take this analysis step by step:

1. Level areas, slopes, and types of soil will control what you use for your watering.
2. Overhead watering, using sprinklers verses ground watering, using soakers, will depend upon your air movement and, again, soil types, slopes, etc.
3. Hard to water areas, some containers, plants, and narrow areas may require certain watering tools.

Overhead watering:

Do you remember when we were told not to water overhead: It brought on diseases, wasted water and the water did not go in deep enough. This is true if you are a "sprinkler". This individual goes out on his lawn in the evening and squirts the lawn till the water runs off, or stands over the flowers and does the same thing. The word "sprinkling" is fine for the seed bed, but for watering, the term is out. We can think of overhead irrigation. It is now done in grape vineyards, cotton fields, grain fields, avocadoes, and all phases of watering. In some flower conditions, the overhead watering may turn to trench watering, due to bloom damage, but to get them started, like in huge gladiolus fields, overhead irrigation takes place. And we all know the watering that takes place on our favorite golf course, freeway plantings, and so on.

1. *Set sieve type sprinklers:* These are our typical portable sprinklers in a variety of types. Some spray square; some spray one area as a fan; some spray round, etc. These tend to drop water in a hurry and may cause quick run off if soil is not in the right condition. They are great for soaking dikes areas where water

can be contained for deep watering. Some of the set sprinklers have a process in dropping drops of water in a patterned manner by rotating baffles and allow for an alternate drop of water, allowing for better water penetration. All of these fit directly at the end of the hose.

2. *Oscillating type:* A number of years ago these sprinklers were developed to overcome the water run off and to provide deeper watering. They arc back and forth, letting water soak in while they go to another area to spread water. They again work in proportion to how your soil can absorb water. The idea is fine. The problem sometimes in water pressure, or the moving parts need servicing and no one to do it! As the years go by, there is no doubt they will improve. They are fine for lawn areas. They fit directly at the end of the hose and can be moved as needed.
3. *The machine gun type (i.e., Rainbird, Rain Jet):* These are the type most commonly found commercially on risers from 3 feet to whatever height is needed. We call it a machine gun because it shoots out very rapid drops of water, with ranges varying from a few feet to many yards. Normally, they throw large drops of water and can be adjusted to spray finer applications. This is an adjusted instrument that is most unusual in what it can do. Several companies make this waterer. You can get it on a portable base. You can buy it on a spike so you can move it as you wish. You can install it in place of sprinkler heads. They have attachments so you can take them in and out to prevent robbery! These are normally used over very wide areas and are not too practical for backyard gardens. You should carefully study their adjustments so you know what they are doing. Frequently, as I travel the freeways and see these in use, some pouring straight streams and wasting "my water", I get fed up. I get out and look at them and they have been bumped or they have lost their adjustments and on and on and on. I am not too sure that the one who does the watering checks to see how they work in the first place. Nuff said! Anyhow, for the farmer, this type of waterer is a godsend for

sure. In using these, you will find they work better for you if you will give some thought to your soil type and water absorption. It might be that they should be turned off at a certain period of time, letting water go in, and then be turned on again. Your soil type certainly governs the kind you are going to buy.

4. *Spinning types:* We have all kinds of these, for sure. Some spin on a little water pressure; some won't work until they have a full force of water. Some can be wheeled across the watering area as needed. These are normally used for lawn areas.
5. *Soakers:* This includes some older types of watering equipment and some new items. These are used at soil level and are made of materials from canvas through plastic, and items to put on the end of the hose for soaking and to prevent washing. Let's review their uses:

   a. *Canvas tube:* This is one of the oldest. The new ones are mildew treated and, if hung to dry after use, last a long time. They can be circled around a fruit tree, put down a furrow, used in and out of the rose patch, etc. The idea is that the water oozes out of the pores of the cloth, slow and easy. They work beautifully.
   b. *Plastic tubing:* These have little holes in them and, when turned on, make a very fine spray 8 to 12 inches high. They can be turned lower to soak, and are called soaking tubes. I find the old canvas much better because after a year in the smog belt, the plastic begins to break and deteriorate. Admittedly, they are smaller in size and blend into the foliage.
   c. *Metal or plastic soakers that screw on the end of the hose:* These baffle water and prevent wash. They can be used for ditch irrigation, basin soaking, etc.
   d. *Wand type soaking tools:* One is a distinct wand about 4 feet long with baffle plates at the end to prevent washing. The other has a slight curve and detachable head that can be used directly at the end of the hose. Both are great for pot

and basket watering, planter watering, nursery watering, and such. They give the effect of breaking down the water into a non-washing flow.

e. *Bubblers:* These are used as inserts into a pipe and water bubbles out, giving a very definite soaking action. Great for installation in the shade garden, in planter boxes and hard water spots, they could even be used as a portable watering instrument in ten foot lengths of pipe in the vegetable garden. They are used commercially for ground watering in contained areas around trees in orchards, etc.

f. *Spaghetti tubing:* At the moment, this has been somewhat limited, for home gardeners to kits, Several have come on the market and were flops. Perhaps the public is not ready for their use. They have been used commercially for years and most of the larger nurseries use them in watering containers. A little, thin piece of plastic tubing runs from a plastic pipe into each pot or given area. This is being used now in the grape vineyards, orchard watering and some large flower and vegetable areas.

Water saving:

Re-read the chapter on mulching.

Watering future:

As water gets even more scarce and water bills go up, we will have watering controls that will run off instruments called tensiomoters that will turn water on as needed, by signal from the soil rather than by a timer or by guess. It is already being done but has not got to the home gardener as yet. Since we are the greatest wasters of water, this is bound to happen. I would rather have it happen now for the pure health of the plants in all our various soil types, and it sure would make my water bill cheaper! I doubt we will ever do too much underground watering because of the eternal problem of alkaline salts clogging holes,, and holes being clogged by minerals of one kind or another. There have been some successful installations but they are few and far between.

How to tell when to water:

This gets down to the real problem for sure. It's one of the most commonly asked questions to any garden authority, nurseryman, etc. It's important to know how water flows through our soil. One reason we condition it is to get better water control, but how do you tell when to water again? Check these for sure:

1. Learn to reach into the soil and look. You can do this with trowel, shovel, soil tube and even a finger in your house plant pots.
2. Beware of plants getting completely dry before you water! Better water when they reach a wrung out dishrag condition.
3. Never keep any plant soaking wet.
4. Factors in watering are air circulation, sun reflection and, of course, your own soil type. Therefore, never set a set schedule. Check frequently as to what's going on in the root area, and don't advise the neighbor how often to water. Tell them to test as you do!

Testing for water:

1. The oldest method of testing for water was digging with a shovel. Then, secondly, was the use of an auger. Both methods are still used. If you can find a soil auger, you have a jewel! I use both. You can even use an old bit, one inch wide, and drill it into the soil with your brace to find out the amount of water in the lawn, pots, tubs, etc.
2. In the last fifty years, we have used what was called soil tubes. These were often home-made and in the last fifteen years are made with stainless steel and have become very practical—and expensive. They run in all lengths, depending on the crop you grow. Some can be pushed into depths of 12 feet. When they are pulled out of the soil, they have an area where the soil is held which you look at as to moisture They run from foot long ones to two foot long ones for average use. A greenskeeper on the golf course would not be without one.
3. Commercially, for years, we have had tensiometers which

measure accurately soil moisture. They are read by a dial and they are expensive. They are probably too expensive for home garden use. Various watering gadgets during the last ten years have come and gone. I have used everyone of them and find that something happens to them along the line, and my money has been wasted. So far, there are two that run on batteries, selling for around $15.00, which do carry good guarantees. I am fearful they will never be to popular until some real demonstrations are done with them for home gardeners.

4. So summarize testing for water, the best still is to look. I will never forget my old English gardener who grew gorgeous flowers in pots. He would never water til he knocked the pot and looked at the soil. He didn't even believe in the "knocking the pot for sound" theory. Maybe he was right after all—LOOK AND FEEL!

FINAL REMARKS

Looking over the final proofs, that means copy as it goes into book form, I found a lot of places I would like to have added a bit more . . . no doubt a year from now, I would like to rewrite the whole thing . . . this is the future of our soil work. We have barely begun to explain how important it is. How basic it all is to our everyday living.

Someday, we will have tools to make deep digging easier. We will have materials to break hard pans faster. We will have a lot more research on container growing for apartment house dwellers. We will have some real down to earth, easy to use, and inexpensive growing techniques of certain things in hydroculture.

We will all benefit in the future as we work with soils a bit more intelligently. I look to a time when we will have great movements back to the acre and its complete living unit. We will profit from our gardening activities physically—and be told by a doctor to do it . . . we will learn more of the mental help in getting rid of worries, for no one worries as he works in the garden and with soil.

. . . And as my mother wrote so much, as we work with the beauty around us, it will rub off on us and make us better individuals. You cannot work with earth and not realize the value of little things . . . that is chiefly what soil conditioning is all about. In daily life, it's the little things that count. . . .

There is a tremendous future for anyone who works with soil professionally, who teaches it, who writes about it, who works with it in his pots, in his backyard, and in his one acre.

EARTHWORMS FOR ECOLOGY AND PROFIT, Vol. 1, by Ronald E. Gaddie, Sr., and Donald Douglas.
The most comprehensive publication on earthworms and scientific earthworm farming available today. Contains details and information which have not been available to earthworm growers in the past. Reveals the "secrets" that books and growers won't tell; all available markets, planning and installing the earthworm farm, beds and bedding, includes soilless bedding, feeds and feeding methods, how to care for earthworms, how to double the size of earthworms in two weeks. Over 175 pages, completely illustrated. **Soft Cover $4.95**
Hard Cover $9.95

VEGETABLE GARDENING KNOW-HOW, featuring Dick Raymond.
One of the best-selling books now available on vegetable gardening, this volume answers hundreds of questions with common-sense, easy, safe, and proven techniques. The generous use of photographs and line drawings throughout make each discussion crystal clear. Shows how to grow 4 to 7 times the amount from the same space with less work. **$4.95**

WHAT EVERY GARDENER SHOULD KNOW ABOUT EARTHWORMS, by Henry Hopp, Ph.D.
Based on years of research conducted by Dr. Hopp and his associates for the U.S. Dept. of Agriculture, this little volume describes for the layman, the many benefits earthworms bring to garden soil. Dr. Hopp describes earthworms' effects on soil moisture, aeration, nutrients, and crop yields, and tells how to increase the worm population in your garden. **Soft Cover $2.50**

THE BUG BOOK, Harmless Insect Controls by Helen & John Philbrick.
A practical and very readable book for easy non-toxic control of common garden bugs. You'll find over 100 common garden bugs clearly illustrated, with complete description, preventive measures and natural remedies. All gardeners, not just "organic" gardeners, will rejoice over the many simple remedies, often using household items. The only book of its kind! **$3.95**

VEGETABLE GARDEN HANDBOOK, by Roger Griffith.
This is a book to be used—red, written in, reread, referenced and brought right into the garden with you. It is a planner's guide, planter's handbook, and gardener's record book all in one. Contains a wealth of information on 57 vegetable varieties to help you decide on seeds, companion plantings, and every other important decision. **$3.95**

LET IT ROT! by Stu Campbell.
This is the home gardener's complete guide to composting. Written in practical, how-to terms, telling exactly what you need to know about composting so you can do it! Tells what materials to use, and what to avoid, good locations, activators, modern applications. Learn here the full, easily understood scientific explanation, and practical procedures for the home gardener. **$3.95**

12,000 COMMON PLANT NAMES by Winton Frey.
If you're a serious gardener, this is a book for you. Compiled by a professor of ornamental horticulture at California Polytechnic State University at San Luis Obispo, this is an index of 12,000 plants listed alphabetically by their common names, showing the botanical name for each. A handy reference for the student or hobbeyist. **$5.95**

CASH FROM YOUR GARDEN by David W. Lynch.
Tells how to sell your own produce at a roadside farm stand. Gives complete details on buying and selling at the edge of the road, initial financing and location considerations, planning the market, market management, advertising, promoting your market, specialties and additional services. Fully illustrated, sure to put money in your pocket.
$3.95

HARNESSING THE EARTHWORM by Thomas J. Barrett.
The classic work on using earthworms for soil improvement and gardening, first published in 1942; now in a new edition with introduction by Ronald Gaddie, author of *Earthworms for Ecology and Profit.* Thomas Barrett combines in this work a wealth of practical knowledge and scientific research, with a deep appreciation for the unity of life.
Contains 166 pages. **Soft Cover $4.95**
Hard Cover $9.95

THE MULCH BOOK by Stu Campbell.
Mulching is the good gardener's secret which saves work, strain, and produces healthy delicious vegetables. Here you'll learn every mulching material from bark to newspaper to stones, mulching perennials and annuals, assisting "finicky" plants, free mulches, exactly what to do when. A complete guide to mulching for the family food gardener.
$5.95

IMPROVING GARDEN SOIL WITH GREEN MANURES by Richard Alther and R. Raymond.
A unique and timely subject for the home gardener in its money saving benefits. Green manuring is the growing and turning under of foliage crops for their fertilizing and soil conditioning values. Clear step-by-step sowing instructions, detailed how-to information on improving problem soils, material and source lists. **$2.50**

DARWIN ON EARTHWORMS by Charles Darwin.
The most quoted book on earthworms ever written. The product of Darwin's life-long fascination with earthworms and their role in the world, now in a new edition with an introduction by Dr. James P. Martin, Professor of Soil Science at the University of California, commenting on Darwin's pioneering insights in the light of modern scientific findings. Contains 160 pages. **Soft Cover $4.95**
Hard Cover $9.95

ORDER FORM (Books)

This order form is for your convenience in ordering books. All books will be mailed promptly on receipt of price and handling fee. Please order, with your remittance, from the same source through which you purchased this book. If your dealer is out of stock you may order direct from Bookworm Publishing Company, P.O. Box 655, Ontario, California 91761.

Note: Prices subject to change without notice.

Gentlemen: Date: ____________________

Please send me the books I have listed below, for which I enclose:

☐ Cash ☐ Check ☐ Money Order in the amount of $ ____________________

Add State Sales Tax where applicable.

Qty. Ordered	DESCRIPTION	Unit Price	Total Price
	EARTHWORMS FOR ECOLOGY and PROFIT VOL. I Gaddie and Douglas — Soft Cover / Hard Cover	$4.95 / $9.95	
	VEGETABLE GARDENING KNOW-HOW Raymond, Dick	$4.95	
	WHAT EVERY GARDENER SHOULD KNOW ABOUT EARTHWORMS Hopp, Henry, Ph.D.	$2.50	
	THE BUG BOOK Philbrick, Helen & John	$3.95	
	VEGETABLE GARDEN HANDBOOK Griffith, Roger	$3.95	
	LET IT ROT Campbell, Stu	$3.95	
	12,000 COMMON PLANT NAMES Frey, Wintor	$5.95	
	CASH FROM YOUR GARDEN Lynch, David W.	$3.95	
	HARNESSING THE EARTHWORM Barrett, Thomas J. — Soft Cover / Hard Cover	$4.95 / $9.95	
	THE MULCH BOOK Campbell, Stu	$5.95	
	IMPROVING GARDEN SOIL WITH GREEN MANURES Alther, Richard & Raymond, R.	$2.50	
	DARWIN ON EARTHWORMS Darwin, Charles — Soft Cover / Hard Cover	$4.95 / $9.95	
	THE COMPLETE GUIDE TO GROWING BERRIES AND GRAPES Riotte, Louise	$3.95	
	BEGONIAS for BEGINNERS Haring, Elda — Soft Cover / Hard Cover	$6.95 / $11.95	
	YOUR ENERGY-EFFICIENT HOUSE Adams, Anthony	$4.95	
	HOME DRYING VEGETABLES, FRUITS & HERBS Hobson, Phyllis	$2.95	
	HOW TO EARN A LIVING IN THE COUNTRY WITHOUT FARMING Osgood, William	$5.95	
		Sub Total	
		Shipping & Handling	.75c
		State tax (where applicable)	
		TOTAL	

Name ____________________

Street or RFD ____________________

City ____________________ State __________ Zip __________

ORDER FORM (Books)

This order form is for your convenience in ordering books. All books will be mailed promptly on receipt of price and handling fee. Please order, with your remittance from the same source through which you purchased this book. If your dealer is ou of stock you may order direct from Bookworm Publishing Company, P.O. Box 655 Ontario, California 91761.

Note: Prices subject to change without notice.

Gentlemen: Date: ____________________

Please send me the books I have listed below, for which I enclose:

☐ Cash ☐ Check ☐ Money Order in the amount of $ ____________________

Add State Sales Tax where applicable.

Qty. Ordered	DESCRIPTION	Unit Price	Total Price
	EARTHWORMS FOR ECOLOGY and PROFIT VOL. I Gaddie and Douglas Soft Cover / Hard Cover	$4.95 / $9.95	
	VEGETABLE GARDENING KNOW-HOW Raymond, Dick	$4.95	
	WHAT EVERY GARDENER SHOULD KNOW ABOUT EARTHWORMS Hopp, Henry, Ph.D.	$2.50	
	THE BUG BOOK Philbrick, Helen & John	$3.95	
	VEGETABLE GARDEN HANDBOOK Griffith, Roger	$3.95	
	LET IT ROT Campbell, Stu	$3.95	
	12,000 COMMON PLANT NAMES Frey, Wintor	$5.95	
	CASH FROM YOUR GARDEN Lynch, David W.	$3.95	
	HARNESSING THE EARTHWORM Barrett, Thomas J. Soft Cover / Hard Cover	$4.95 / $9.95	
	THE MULCH BOOK Campbell, Stu	$5.95	
	IMPROVING GARDEN SOIL WITH GREEN MANURES Alther, Richard & Raymond, R.	$2.50	
	DARWIN ON EARTHWORMS Darwin, Charles Soft Cover / Hard Cover	$4.95 / $9.95	
	THE COMPLETE GUIDE TO GROWING BERRIES AND GRAPES Riotte, Louise	$3.95	
	BEGONIAS for BEGINNERS Haring, Elda Soft Cover / Hard Cover	$6.95 / $11.95	
	YOUR ENERGY-EFFICIENT HOUSE Adams, Anthony	$4.95	
	HOME DRYING VEGETABLES, FRUITS & HERBS Hobson, Phyllis	$2.95	
	HOW TO EARN A LIVING IN THE COUNTRY WITHOUT FARMING Osgood, William	$5.95	
	Sub Total		
	Shipping & Handling		.75c
	State tax (where applicable)		
	TOTAL		

Name ____________________

Street or RFD ____________________

City ____________________ State __________ Zip __________